Grant Fowlds is a South African conservationist with a unique commitment to everything endangered. Grant's vision is to try to fill the shoes of the late elephant whisperer Lawrence Anthony and his recent undercover filming of a tusk and horn auction in Vietnam demonstrates that he is on the right track.

Graham Spence is a journalist and editor. Originally from South Africa, he lives in England. Together he and his brother-in-law, conservationist Lawrence Anthony, wrote *The Elephant Whisperer*, the story of the incredible relationship forged between one man and a herd of wild African elephants. Other books with Lawrence Anthony include *Babylon's Ark: The Incredible Wartime Rescue of the Baghdad Zoo*, and *The Last Rhinos: The Powerful Story of One Man's Battle to Save a Species*.

D1157235

Grant Fowlds and Graham Spence

SAVING THE LAST RHINOS

One Man's Fight to Save Africa's
Endangered Animals

ROBINSON

ROBINSON

First published in Great Britain in 2019 by Robinson

This paperback edition published in 2021 by Robinson

Copyright © Grant Fowlds and Graham Spence, 2019

1 3 5 7 9 10 8 6 4 2

The moral rights of the authors have been asserted.

A CIP catalogue record for this book is
available from the British Library

ISBN: 978-1-47214-251-1

Typeset in Caslon Pro by SX Composing DTP, Rayleigh, Essex
Printed and bound in Great Britain by Clays Ltd, Elcograf S.p.A.

Papers used by Robinson are from well-managed forests
and other responsible sources

Robinson
An imprint of
Little, Brown Book Group
Carmelite House
50 Victoria Embankment
London EC4Y 0DZ

An Hachette UK Company
www.hachette.co.uk

www.littlebrown.co.uk

This book is dedicated to the more than one thousand rangers who have given their lives on the front line in the ongoing battle to save endangered wildlife in Africa. Forever remembered.

On a more personal level, thank you to my wife, my three girls, my parents and my dedicated brother, without whom this book would not have been written.

Contents

Prologue

The one thing I dared not do was what I felt like doing most. Breaking out in a cold sweat.

If I had, it would have given the game away.

In my laptop bag were two packets of shaved rhinoceros horn. Two plastic Ziploc bags, the kind corner-shopkeepers use when they bank a day's cash takings. Ounce for ounce, the shavings were as valuable as gold, and far more valuable than drugs. If caught, my fate would be similar to a common cocaine smuggler with the likelihood of a lengthy jail sentence and six-figure fine.

I was at the impressively modern Noi Bai airport, better known as the Hanoi International, in Vietnam's capital city. I was about to board a plane for Johannesburg, South Africa, after one of the most invigorating and uplifting experiences of my life in conservation, attending the 2015 Operation Game Change festival, WildFest. It was the first to be held in a country regarded as the world's most notorious wildlife crime clearing house, and the fact that Vietnam was the venue was a significant breakthrough in the battle for the planet.

The first part of the week, I witnessed animated young

people singing, dancing, screening anti-wildlife-crime films and expressing hope for the earth.

I was filled with optimism.

Then it all came crashing down. I saw for myself the dark flipside of the coin.

It was at a village called Nhi Khe, billed in tourist brochures as a quaint street-market centre about 11 miles outside Hanoi, ostensibly dealing in traditional Vietnamese crafts.

The reality is far more sinister. For a conservationist, this is the most evil square mile in the world; the global trade 'capital' of acutely endangered wildlife. It is estimated that 50 per cent of all threatened-species products are sold at Nhi Khe.

It was there that I bought the highly illegal rhino-horn shavings, which was the reason for my imminent cold-sweat panic as I neared the airport's X-ray machines.

A few days previously, four of us had gone to the so-called craft village to verify first-hand if rhino horn was being openly traded on the streets. With me were my conservation colleague and friend Richard Mabanga, Johannesburg photographer Ilan Ossendryver and Jim Ries from the American NGO One More Generation.

As we walked down the maze of alleyways housing Nhi Khe's cramped sidewalk shops, looking starkly conspicuous as foreigners, two shady modern-day spivs sped down the street on motorbikes, shouting in Vietnamese and beeping their horns.

It was obviously a warning and the wooden shutters of the shopfronts started slamming as we approached. Thud-thud-thud . . . like clappers beating the bushveld with sticks to flush out buck from dense foliage during a hunt. I noticed that not all shops were closing, probably just the nastier ones. Those stocking rhino horn, ivory, pangolin and tiger products under their counters. It was incredible to think that this pokey, narrow street had millions of dollars of contraband for sale. A full rhino horn sells for more than a top-carat diamond.

We were technically undercover but Richard, Ilan and I must have stuck out like nudists in a cathedral. The last thing we wanted was to be identified as Africans, the home of the animal horn we were seeking, so we concocted a flimsy cover with me and Ilan coming from Europe and Richard a Jamaican. He even had a go at the accent – it was pretty clear that Bob Marley was his inspiration. All the same, it was far more convincing than me trying to speak Queen's English with my South African twang.

I acted like a casual tourist, even wearing a cheap 'I love Vietnam' T-shirt, nonchalantly inspecting the vast array of goods on display. Much of what I saw broke my heart. While I didn't spot rhino horn, there was everything else imaginable, including ivory, tiger parts, tortoise shells and pangolin scales sold as carelessly as loaves of bread.

To boost my tourist credentials, I bought some rosewood souvenirs – the most trafficked and endangered hardwood in the world. Every time I mentioned 'rhino horn' to a shopkeeper there was a curt shake of the head.

Then suddenly we got a bite. At one shop a man caught my eye and beckoned. Four other men were with him. Yes, he said. They had horn.

We were taken inside and in an instant the mood changed. They started searching us, barking aggressively in broken English. Who were we? What were we doing in Nhi Khe? Why did we want rhino horn?

I put my hands up, talking quietly while trying to defuse the situation. I said I was suffering from a cold I could not shake and wanted to try a traditional remedy.

It was the right move. They instantly calmed and stopped shouting.

One then smiled and put out his hand. I shook it. He leant forward, almost whispering in my ear.

'How much money do you have?'

We had previously scraped together $500, all of us pooling our resources. This was way off-budget for the conservation charities I work with, as I couldn't really put 'rhino-horn purchase' on an expense account. Rhino charities are not in the habit of buying rhino horn. In fact, the extreme opposite. We were funding this operation ourselves.

Five hundred dollars is serious money in dong, the local currency, but the dealers were unimpressed. I could sense their expectations deflate like a pricked balloon. The going price for horn was $16,000 a kilogram. One of the men said dismissively that $500 would buy a paltry 32 grams of shavings.

I nodded, saying I only wanted a small amount for my cold, and handed over the money. In return I got two small Ziploc bags filled with what looked like bone splinters.

Back in our hotel, I held the packets in my hand. Some magnificent beast died for this pathetic amount, sold to cure a common cold. A towering icon of the animal kingdom, a species that has survived the dinosaurs, ice age, meteor strikes and every other catastrophe except man, had its horn hacked off and been left to bleed out in agony in the bush so some human could stop sneezing. And it doesn't even work. A rhino horn is not medicine; it is keratin, the same stuff as our fingernails. The indignity of human greed and ignorance is infinite.

However, I was not convinced this was genuine rhino horn. I saw so many Asian buffalo products on the glass-top shop counters in those cramped Nhi Khe streets that I was certain the packets contained nothing more than shredded bovine bone. I reckoned they knew we were small-time amateurs and had taken us for a ride.

There was only one way to find out. I needed to get the packets back to South Africa and have the contents tested. If they were rhino shavings, the horn would be from Africa.

In other words, I had to smuggle the horn home.

The risks were daunting. If caught, I would be thrown into jail and hit with a crippling fine that I could never pay on my NGO salary. I had a wife and three daughters at home, who would not be thrilled with the idea of their dad locked up in some dingy, far-flung cell.

However, there was no question in my mind. I knew I had to see this through. If I didn't, it would be tantamount to making a break in a rugby match with a clear run for the try line and then kicking the ball out of play. The big question was not whether I did it, but how.

This was trickier than it may sound as even though Vietnam at the time had an unfortunate reputation of turning a blind eye to wildlife smuggling, the country was belatedly starting to clamp down thanks to international pressure. It would be just my luck that one of the first 'traffickers' the police caught was a conservationist doing freelance undercover work.

I had three smuggling options. I could either keep the packets on me and take my chances with the walk-through electronic scanners; slip them in my laptop case and pray they weren't picked up by the X-ray machines; or else pack them in my suitcase going into the aircraft's hold.

A packet with some bone-like shavings might possibly – just possibly – not spark too much interest when viewed through an X-ray scanner, but a sniffer dog would go ballistic if it got the scent from a suitcase. I reckoned it would probably be safer to carry the packets in my hand luggage.

But I was guessing. I had no idea. The bottom line was that any option was just winging it and hoping for the best.

I was inwardly nervous but somehow outwardly calm as I walked through the electronic scanner. I half expected a police officer to cuff my wrist and haul me away, but to my relief the laptop bag came bumping down the conveyor belt from the X-ray machine without any eyebrows lifted.

I breathed a little easier once on the plane, and a cold beer from the drinks trolley was just the medicine I needed. Unlike rhino horn, it worked.

In South Africa I was banking on the fact that few, if any, customs officials would think someone would be crazy enough to smuggle rhino horn into the country of origin. It would be tantamount to sneaking cocaine into Colombia. But had any-one told the sniffer dogs that?

I thought of what I would say to my wife Angela if caught. It was a waste of time and mental energy, as there would have been nothing I could say. I was returning from a trip to a festival spotlighting wildlife crime, and here I was in the eyes of the law committing exactly that. How could anyone explain that away?

Looking as cool as possible, I got through the 'nothing to declare' queue at O.R. Tambo airport in Johannesburg. Once out of the building, I breathed deeply, savouring the crisp Highveld air of my home country. I had done it. And I was still a free man.

The next step was to deliver the samples to Dr Cindy Harper, the General Manager at the Onderstepoort Faculty of Veterinary Science at the University of Pretoria. Cindy is truly one of wildlife's unsung heroes, working tirelessly to identify the scores of rhino horns being smuggled to the Far East. But I could not compromise her with smuggled shavings, so didn't tell her where I got the samples from. Instead I filed a request for a DNA test on samples from an 'unknown location'.

Two weeks later I got the results. It was not Asian buffalo, as I first suspected. The dealers were not con artists fleecing gullible tourists. This was the real deal; genuine rhino horn. Cindy could not pinpoint exactly where it was from, but the DNA evidence was otherwise conclusive. The shavings came off a southern white rhino from somewhere in South Africa.

I whistled softly. Not only had I concrete proof that South African horn was being sold openly in Vietnam, I'd shown how easy it was to smuggle it through a sophisticated airport with some of the most modern scanners and X-ray machines available.

I felt sick to my stomach, yet it was a moment of cold clarity. The eco-wars were to become the focus of my life.

CHAPTER ONE:

Goats and Growing Up

Goats are not the most glamorous of animals.

They are not cute or cuddly like kittens, they don't obey commands like dogs, and they are not sleek and beautiful like lions or leopards.

But I have huge affection for them. Not only are they among the most indomitable of creatures, goats were the start of my love affair with animals and the magnificent wild places of Africa.

I was born on a 2200-hectare farm called Leeuwenbosch outside Port Elizabeth in the Eastern Cape province of South Africa. It was a sheep and cattle ranch which had been owned by the Fowlds family since 1872, and much of the land was still untamed. Aloes stood tall like spires of stately cathedrals, while grasslands and savannah spread lush-green in the valley of the Bushman's River that snaked through our lands on its way to the sea. On the higher ground, shrub thickets and thorn trees with spikes like daggers dotted the terrain, sprawling up the rocky kopjes – the hillocks – as they had since time immemorial.

It was a childhood with unimaginable freedoms compared to today. Like many other white boys growing up in southern Africa, my first language was Xhosa and my playmates black kids living on the farm. When I was six or seven, I would often disappear into the bush for several days, and after a night or two my mother would start worrying and ask the workers if they knew where I was.

The answer was always the same. Yes, I was fine and had just eaten breakfast of *umphokoqo* – dried, crumbly maize meal sometimes served with sour milk called *amasi* – at the staff huts. To this day, I would choose *umphokoqo*, the Xhosa staple diet, over any other breakfast, although it is equally delicious for lunch or dinner.

Mom would then despatch a search party to bring me home for more food and wash my ears, which were so filthy she could have planted potatoes in them.

I was the oldest child, born a year before my sister Ros. We had a younger sister, Mary-Nan, who drowned in a reservoir when she was three. My father was away playing cricket at the time, and the anguish that caused my parents was incalculable. A decade was to pass before they had children again; my brother William and, a year later, Jayne. Both Mom and Dad came from rugged pioneer stock and were tough people, but they were always gentle with us. My dad once said to me that parents who are hard on their children have never experienced the loss of one.

Despite that period of almost bottomless grief, Leeuwenbosch thrived as a farm. There had been tragedies before Mary-Nan, as the unusual appearance of four cypresses outside my parents' home attests. The trees are the sole alien species in an indigenous landscape. They were planted by my great-grandparents, William and Gertrude Fowlds, in memory of the four children they lost in a row as infants early last century. As my father wrote in his memoirs, 'One is astounded to think how parents could cope with such a tragedy.'

Cope they did. The people of the Eastern Cape district of Alexandria are mainly descendants of the resilient British settlers from the 1820s onwards and the impressive Xhosa tribes with whom they clashed during a hundred-year-long border war. There is much blood in our history, much tragedy, much courage and much goodness. I was privileged to grow up there.

The defining moment of my childhood was when my maternal grandfather, Claude Rippon, took me to a trading post at the Carlisle Bridge on the Great Fish River one weekend. I still have no idea why, as Carlisle Bridge is one of the most barren places in the country. The joke is that when a plague of locusts swarmed, looking for countryside to strip bare, they bypassed Carlisle Bridge figuring they had already been there.

We passed a yard where goats were sold. I even remember the name of the breeder, Mr Norton. For some reason, I decided there and then that I was going to became a goat farmer. With the dogged persistence of youth, I persuaded my grandfather that I simply could not live without the animals.

What he made of a seven-year-old wanting to buy semi-wild goats, I have no idea, apart from scratching his head and assuming I was a little unusual. Or, more likely, crazy. But grandfather Claude was the kindest man I have ever known and he bought me five ewes and a ram.

A few days later I returned home from school and was told that 'the present' from my grandfather had arrived. I can't remember ever being so excited. My life changed on that day.

3

The goats were as skittish as wild horses, which came as a surprise to me as I was expecting something tamer, like sheep. I was rudely disabused of that notion as we tried to corral them into a nearby camp. Fortunately, the workers on Leeuwenbosch were experienced stockmen and somehow we coaxed the wild, rangy animals into a grazing enclosure.

My new life began. Soon I was breeding the animals, and my dad seconded one of his workers called Tolly Masumpa to help me.

Tolly was a godsend. He had been badly burnt in a veld fire some years ago when he tried to kick a can of paraffin out of the way and the flames flared up his legs. His limbs were covered with pink scar tissue and, as a result, he couldn't do the hard manual labour that mainstream farm work required. He also couldn't ride any more, which was a problem as this was before the time of off-road motorbikes and most stock herding was done on horseback.

But Tolly was undefeatable. Being lame was just a minor hassle as far as he was concerned, and we bought him a Scotch cart – a two-wheel cart – at an old farm auction with a drawing horse to help him get around.

Every day Tolly would ride to the camps in his cart and we would round up the goats if they needed to be herded to new grazing, or for dipping. He even had a compartment in the back of his cart where he put the kids that couldn't keep up with their mothers.

Boer goats are extremely hardy creatures, requiring very little maintenance. They only need to be dipped for lice and ticks every week or so. Other than that, they fended for themselves, but I always kept them in camps close to our house where I could be with them. I was so fanatical that my dad eventually told me to stop physically handling them. It worried him that I loved the animals too much.

The herd was expanding nicely and I persuaded my dad that we needed to increase our stock at an even faster rate. We went to an auction and bought the mangiest black and blue goats we could find. We only paid a pittance for them, but they were pitiful specimens. However, that was my plan; to put decent rams on substandard ewes, which both accelerated the growth of the herd and upgraded bloodlines. Each generation would produce better-quality animals.

Goats were the focal point of my life. So much so that I was called 'Goat' at school, which wasn't the coolest nickname to have. But it gave me some minor celebrity status as every Wednesday afternoon I got time off to attend auctions and buy and sell animals while other kids were playing rugby or cricket. I even got a write-up in the local newspaper, which in those days was the street-cred equivalent of getting a hundred thousand likes on Facebook.

By the time I was sixteen, I was running more than a thousand animals. The profits paid for my school fees, which was no mean feat seeing I was at St Andrew's College in Grahamstown, an expensive private school.

I couldn't have done it without Tolly. He was brilliant. He could not read or write – in fact, he could only count to five – and yet he knew every animal by sight. At times we would be driving a herd of more than four hundred to the dip and he'd tell me one was missing. I would say no, they're all there. He would shake his head, muttering that the one that always walks at the back, or the one with the red patch, or the one with the bent horn, was not in the herd.

We'd go back and search the camps and, without fail, Tolly

was right. We always found the goat he had described, whether it was lost, sick or, sadly, dead.

I still find it incredible to think that the entire herd of unruly creatures was run by myself, barely a teenager, Tolly, who was lame, and my border collie Kola, who slept by my bed each night. Kola had never been formally trained, yet he could corral the animals as expertly as any of the top sheepdogs in Britain.

Despite his injuries, Tolly lived until he was eighty, dying soon after my fiftieth birthday. He was one of the best; my finest mentor and friend, even with our age difference.

What I didn't realise at the time was how much dealing with goats drew me deep into the ancient spirituality of Xhosa culture. A goat to both Zulu and Xhosa tribes is more than just a sturdy animal uniquely adapted to the harshness of this continent. It is a mystical creature used in rituals and ceremonies. Ancestors are extremely important in all African cultures, and almost all my buyers were black South Africans wanting quality goats to commune with the spirits of their forebears through sacrificial slaughter.

I did on occasion send my animals to the abattoir, but was so upset at seeing them driven into the slaughterhouse that I gave it up. The slaughterhouses also paid peanuts. A Xhosa wanting a goat as *lobola*, or bride-price, to impress a future father-in-law paid far more than any abattoir catering for the white market.

Even that was hard for me. I hated to see my animals tethered by the neck and led off down the road. I had constantly to tell myself that I was a businessman, not a pet breeder.

At that stage, I was also still running wild with Xhosa boys on the farm, hunting, fishing and unravelling mysteries of the bush with barefoot botanists who may not have known any

Latin names, but sure as hell knew how to live off the land. I discovered how to dig deep in the soil for succulent bulbs that could be squashed for drinking water and then the pulp could be eaten. I learned how to pick prickly pears without getting ripped by cactus thorns and how to smoke out a beehive with a match and tattered piece of hessian to get the honey.

We would be gone for days, but this was no camping with fancy tents, fleece-lined sleeping bags and blow-up mattresses. We sheltered in an old rusty water tank with a hole for a door. The days were blazing hot, but the nights often cold with thin frost crisping the ground. We huddled around red-glowing log fires, roasting small animals or birds we had shot for dinner, then creeping exhausted under a shared blanket on the hard earthen floor.

My friends were almost always older than me. Some were *abakhwetha*, teenagers who had just been through a circumcision ceremony known as *Ulwaluko*, which in Xhosa culture is the transition from youth to adult. Sometimes they showed me their recently circumcised manhood, which looked extremely painful. I believed that I too would have to undergo an *Ulwaluko* before becoming a man, and was not looking forward to it. The Xhosa youths laughed when I told them that. *Ulwaluko* was not for white boys, they said.

I also learned how to hunt with my black friends, first with catapults, then, as we got older, with .22s and a .410 shotgun for guinea fowl and other game birds.

But I soon discovered the hard way that my love for animals would kill my love of hunting stone dead.

It was one of the most traumatic and long-lasting lessons of my life.

CHAPTER TWO:

Hunting for the Right Answers

F or those of us living in the outbacks, hunting was wired into our DNA.

Although the area was no longer the frontier – and had not been so for more than a hundred and fifty years – the pioneer spirit still prevailed powerfully. More and more land was being cleared for agriculture, but there was a lot of wild game in the remaining bush and virtually all farmers hunted to some extent.

At Leeuwenbosch, we were no different except that we were possibly more organised than most. On the first of July each year, my dad and his many friends would congregate at a safari camp called Ashcombe to take part in a large hunting festival that became a tradition.

It was started by my grandfather Victor Fowlds and his neighbour Guyborn Slater in 1938. However, in Victor's case it was more for family reasons than hunting as his mother was not well and he didn't want to holiday far from home in case she took a turn for the worse. Instead, he decided to set up a hunting camp at Ashcombe, 7 miles away on the other side of the road, so he could have a break from the long months of farming but still be close by in the event of an emergency.

Initially the hunts lasted for a weekend or two, but the fame of Ashcombe soon spread far and wide as the word got out that this was no ordinary jaunt in the bush. It was more like something out of a Hemingway novel with tents, caravans, long-drop toilets, outdoor showers and log fires with iron grids for grilling slabs of venison. But unlike Hemingway's books of romanticised Africa, this was the real deal with rough-and-ready men clutching frosted beer cans or whisky glasses with ice tinkling and swapping yarns about their escapades of the day. On some nights the entire district would arrive for festivities and music lasting long into the night. It was a big deal to get the nod to go on an Ashcombe hunt and we had people from all around the world joining us. If you declined, you were not asked again. Few declined. This was the safari of the year.

The schedule never varied. Hunting took place on alternate days with long racks of South African jerky called *biltong* being cured on the rest days. My grandfather also insisted on a camp rule that every morning was started with a cold shower, which was something you would not have found on a Hemingway safari where Swahili servants ran hot bath water in canvas tubs for the *bwana* and *memsahib*.

My dad carried on the Ashcombe hunts for another ten years after his father died, and I remember as a boy going on those safaris and loving every minute. However, even in those days, I enjoyed the convivial fireside chats, the stories told by thorn-scratched men of the bush, the lively banter and sheer exuberance of living outdoors far more than the actual hunting.

The hunters often used dogs as much of the growth was so thick as to make tracking almost impossible. The indigenous Eastern Cape bush, called Albany thicket, is among the most

9

impenetrable vegetation in the world. It's a type of dense scrubland consisting of short thorn trees, shrubs and creepers that can reduce visibility to a few yards. In 1919, Major P. J. Pretorius – one of South Africa's most famous hunters whose exploits inspired Wilbur Smith's novel *Shout at the Devil* – was hired by the provincial government to shoot elephant herds in the Addo area to make way for farmland. He described the bush as plucking a tract of the toughest, most torturously tangled jungle in wildest West Africa, and dumping it in the Eastern Cape.

Among the more colourful characters of the Ashcombe safaris was my great-uncle, Dudley Fowlds. On one occasion he was part of a group beating the bush to spook buck into an open area, when something struck him in the neck.

It was painful, but nothing could stop him from the chase and he continued beating until he had finished his section of thicket.

He sat on a rock in the blazing sun and one of the other hunters glanced at him and remarked, 'Dudley, you're looking a little white, *boet*.' (*Boet* is the Afrikaans word for brother, a generic term for friend.)

'*Ja*. I think I've been shot.'

There was blood gushing down his neck.

'How does it feel?'

Dudley touched the wound, looked at the blood on his hand and said, 'It feels a bit warm.'

It then transpired that he had been shot – clipped in the back of his neck from a ricochet accidentally fired by another hunter.

Dudley was driven to the local doctor, who told him he had no anaesthetic.

'Just cut the bullet out, *boet*,' Dudley said. 'I need to get back to Ashcombe.'

Dudley didn't make a murmur as the doctor sliced open his neck with a scalpel and removed the chunk of lead. It missed his spinal column by a fraction of an inch. Otherwise he would have been paralysed. Or dead.

The next day, Dudley was out hunting again.

Such stories were not that unusual in the Eastern Cape fifty years ago, but Dudley was a legend in his own right. An immensely powerful man, he would lift his Morris Minor truck with his hands to put it on blocks and change a tyre rather than bother with a jack. The small truck also had a long whippy aerial bolted onto the middle of the bonnet, but not because Dudley had a fancy radio in his clapped-out jalopy. It was to slice through spider webs in the bush as there was no roof on the vehicle.

He also was 'famous' for falling off the top of a windmill and crashing to the ground without seriously injuring himself. When asked how he survived, he would look at the questioner askance and say, 'Well, *boet*, I fell slowly.'

As well as going on the Ashcombe safaris and hunting birds and small game with my Xhosa friends, sometimes I would go out into the bush to shoot a buck for biltong, usually with an extraordinary guide called Lastnight Masumpa, the brother of my goatherd Tolly. As far as I know, Lastnight was his real name, not some moniker given to him by a non-Xhosa speaker.

He was a maestro in the bush. He was employed at Leeuwenbosch as a gardener, but he had the true heart of a hunter. Hunting was what he lived for. He showed me how to track and stalk, and although I never hunt any more, those skills still come in handy when I'm in the veld doing conservation work around Africa.

However, on what turned out to be my most memorable hunt, Lastnight was not with me. Which is probably why I made an amateur mistake that still haunts me to this day.

I was in my mid-teens and shot and wounded an antelope. It was bleating like a child and I rushed up to it, planning to put it quickly out of its misery with my pocket knife.

I started trying to cut the carotid artery behind the horns that pumps blood to the brain. I thought one stab would end it all.

I was wrong. Horribly wrong. I stabbed and hacked away while the poor animal bleated louder and louder, each bawl sounding more pitiful. It cried like a human baby.

I started calling for help, but was too far from the house for anyone to hear me. Eventually I carried the bleeding creature home and finally managed to kill it.

It was one the most traumatic yet important lessons of my life, but sadly an animal had to die to teach it to me. I now knew without doubt that I wanted to conserve nature rather than destroy it. It was the first tentative step I took to becoming a conservation activist.

I never hunted again.

CHAPTER THREE:

Wild Lesson, Wild Men, Wild Coast

I was not much of an academic at school, and that's putting it politely.

My heart was outdoors. Stuffy classrooms, books and lessons did not agree with me.

The exception was Xhosa. It was a subject at which I excelled. Perhaps this was not surprising as I was already fluent in the language, and as most of my time on the farm was either running goats with Tolly Masumpa or running wild with Xhosa youths, I was completely at home with black South Africans.

However, my college Xhosa teacher was white and one of the biggest influences of my young life. His name was Alistair 'Ali' Weakley and, for me, he was a carbon copy of the inspirational teacher played by Robin Williams in *Dead Poet's Society*.

In those days there were few black teachers in white schools and so it was not unusual to have a white guy teaching the home tongue of Nelson Mandela. But Ali was not your average white guy. Not only did he speak the language beautifully, but he was enthralled by Xhosa culture and traditions.

This was rare in the 1970s, as like most rural areas the Alexandria farming district was conservative, even by South African standards. There were few liberals in the countryside, and Ali was a notable exception. He was so far removed from most other whites of his generation that they could have come from different planets.

But Ali could pull it off like nobody else. He defied every stereotype of a liberal. For a start, he was an excellent rugby player, hard as granite, and could always be found bloodied and bruised at the bottom of a scrum. He was captain of the Border first team, and although Border may not have won a cabinet full of trophies, they were feared throughout the land as the meanest, toughest players on any field.

Ali did not exactly have film-star looks, but he was a babe magnet and seldom seen without a gorgeous girl on his arm. So to us he was as glamorous as a movie star, with wild long hair, flashing a gap-toothed smile and speeding around the town in a purple Volvo.

All the boys worshipped him as he was so unconventional – arriving late for formal school assemblies with uncombed hair and his academic gown hanging untidily off his brawny shoulders. No doubt other teachers hated his nonconformity as much as we loved it.

I got a First Class pass in Xhosa, which for me was an academic miracle as I barely scraped through in other subjects. But I think every boy in Ali's class got an A; his lessons were that good.

He also was one of the few whites who supported Nelson Mandela's African National Congress (ANC), which was banned at the time. But such formalities did not bother Ali. He was a total rebel, and a completely fearless one.

The highlight of my school career was when Ali took us on a ten-day field trip along the Wild Coast, a savagely beautiful stretch of rocky seashore curving down the Transkei, which is what the Xhosa homeland was called in the apartheid era.

We camped on the beaches, hiking into the dunes to visit Xhosa kraals, or homesteads, where Ali showed us first-hand the rich heritage of this proud, impressive people.

I subconsciously knew a lot about Xhosa customs as I mixed so much with them on the farm, but I never really thought about it. Ali brought it all to life, explaining the traditions to us with vivid words and images as we sat around a campsite fire on the beach. Ali loved lots of things in life – rugby, women, fishing, partying – but also teaching. On that trip, he shared his abiding passion for Xhosa culture unstintingly with us.

I remember being severely reprimanded when I entered a kraal uninvited to look at the cattle of the *induna* (headman). I did so because as a farmer I genuinely was interested in cows. We had a large herd at Leeuwenbosch and I was curious to see the condition of the stock.

Ali called me over and gave me one of the biggest tongue-lashings of my life.

'You never ever go into a kraal without asking the *induna*'s permission,' he fumed. 'Only if he invites you in are you allowed to inspect his cows and praise them. It's the same as walking up to a stranger and asking if you can see his bank balance.'

It's a lesson I never forgot.

We also learned about the circumcision ceremonies of boys becoming men, what the ochre they smeared on themselves meant, the symbolism of the different colours that the women wore, how they mash corn – there was no facet of Xhosa life that didn't fascinate Ali. He was so enthused that he carried us along with him by sheer force of character.

I only realised many years later what a masterclass of Africa I was getting from a man who truly was in tune with the soul of the continent. Ali's eccentricity and exuberance opened an exciting new door for me. My knowledge of African languages and culture has been one of the greatest gifts bequeathed to me. It has stood me in good stead throughout the continent. It was a rare privilege, bestowed on me by people such as Ali, which I will always treasure.

I wish this had a happy ending. It doesn't. In 1993, Ali was murdered by the people he loved.

It happened after Chris Hani, the charismatic leader of the South African Communist Party and chief of staff of *uMkhonto we Sizwe*, the armed wing of the African National Congress (ANC), was assassinated. Hani was outside his house in Boksburg near Johannesburg when a Polish anti-communist immigrant named Janusz Waluś shot him as he stepped out of his car.

The country held its breath, waiting for a mass uprising and waves of bloody revenge attacks, if not outright civil war. Nelson Mandela went on TV appealing for calm, showing what a great statesman he was. Mandela pointed out that it was a white woman who called the police, resulting in the prompt arrest of Hani's killer.

The fury in the townships and on the streets was palpable, but most people heeded Mandela's call. Sadly, not all.

Three days later, Ali and his brother Glen were returning from a fishing trip at the Umngazi river mouth. On the road to Port St Johns, four youths armed with automatic rifles opened fire on the vehicle, killing the Weakley brothers.

The pathos was almost too much to bear – a man who had spent his life helping blacks hamstrung by the iniquitous apartheid system being killed by an ANC cadre. The liberation struggle in this case had tragically devoured its own.

Several years later at the Truth and Reconciliation Committee (TRC) amnesty hearings, one of the killers, Mlulamisi Maxhayi, gave his reason for the murder: 'We decided to kill the white people because they were a symbol of apartheid.'

Ali was anything but a symbol of apartheid. He was a symbol of hope. I still shudder when I think of such a monumentally tragic waste, and the terrible loss to our country.

But yet . . . consider this. Fundisile Guleni, another of the killers, told the TRC: 'We are so sorry for the families and would like to apologise to all of those affected by our actions.'

That, I believe, is what Ali would have wanted to hear.

I left school to do mandatory military service, but was lucky enough to be accepted into the Air Force rather than becoming a foot slogger. I was promoted to lieutenant and fell in love with flying, crewing in Mirage and Impala jets as well as Alouette helicopters. As luck would have it, I was based in Hoedspruit, which is in the heart of wildlife country and a gateway town to the Kruger National Park, South Africa's most famous reserve. Even better, my living quarters were in an actual game reserve and I used to jog past herds of buck and zebra every morning.

I briefly toyed with the idea of becoming a full-time pilot, but the lure of farming was too strong. After leaving the Air Force, I enrolled at the Cedara Agricultural College in Pietermaritzburg to study for a farming diploma. But more importantly, it was there that I met my future wife, Angela Townsend, a gorgeous blonde at the nearby Teacher Training College.

In 1985 I returned to Leeuwenbosch, and although I still had my goat herd, I decided to go into dairy farming. My dad agreed with me and, in a strategic family move, I bought from him an adjoining 400-hectare farm called Sunnyside, on the banks of the Bushman's River.

Angela and I married in 1987, and by 1989 I had built a state-of-the-art dairy and was the second largest protein milk producer after DairyBelle, a household name in South Africa.

Soon afterwards, along with a few other farmers in the area, I was approached by a consortium of Johannesburg businessmen and made an offer that seemed almost too good to be true. In essence, they would not only buy our livestock off us, they would then pay us a management fee to graze them.

The first lesson I learned from that deal was that when the suits from Jo'burg knock on your door, you need to scrutinise every word, dot and dash in a contract, no matter how minuscule.

The second is that when an offer seems too good to be true, it usually is.

What I later discovered was that the suits owned all the assets, but as I was the trading company, I still stood surety for bank repayments.

Then one of the worst droughts in living memory struck and the Bushman's River dried up. Technically, the cattle belonged to the suits, so when I started trucking in water and expensive livestock feed, they said the costs were too high and instructed me to slaughter the entire herd. Their philosophy was simple; the first loss is the best loss. They didn't give a damn about starving animals; cut their throats was their answer.

I couldn't do it. I could not kill off a herd of prize Guernsey cows just to balance some city slicker's books. I may be a sentimental farmer, but that was not the way I did things.

Initially I used my own money to feed the herd, but it was eventually out of my hands as the sheriff of the court arrived and impounded the animals. I fought back with an interdict and, with the help of Paul Benham, one of the suits who actually showed some compassion, I managed to save half of the herd. But it crippled me financially and eventually I closed the dairy down.

At that stage Angela's father, Dave Townsend, a major sugarcane grower in KwaZulu-Natal, asked me to come and farm on the North Coast. He obviously wanted his daughter closer to home, but at the same time he was planning on getting out of farming and the fact that I was looking for a job was a fortuitous coincidence.

Angela and I, with our first child Jess, who was now three years old, arrived in KwaZulu-Natal in 1991. My first task was to learn how to grow sugar, which was vastly different to dairy and sheep. I also planted bananas, which was so successful that within a few years they overtook sugar as my main crop.

We were soon up and running. In fact, we were thriving. But at the same time I was now getting more and more interested in conservation, perhaps even at the expense of farming. I remembered how happy I had been in the bush back in the Eastern Cape, the freedom and feral beauty of the wilderness, and somehow knew that was where my true calling lay. I had no concrete ideas of how to channel this vision into a viable project; all I knew was that it was something I wanted to do.

But how?

Not long afterwards I got an answer. In 1991, one of the world's most gifted eco-visionaries, Adrian Gardner, opened Shamwari, a game reserve just to the north of Leeuwenbosch. He had started the project the year before with the humble purchase of a small, badly degraded 1200-hectare farm. The same drought that had crippled me also ruined many other farmers, who were forced to put their land on the market. Adrian, originally from Zimbabwe, bought them out and today the Shamwari Game Reserve is an impressive 25,000 hectares.

Having grown up in the area, I knew first-hand that the Eastern Cape once teemed with game. In terms of biodiversity, it's one of the richest wildlife ranges in Africa. It was here that the Big Five, lion, leopard, buffalo, elephant and rhino, were first encountered by the early European settlers. That's why wildlife historians still refer to the Cape buffalo, Cape leopard and Cape lion. The endangered black rhino also once flourished.

But due to hunting, chronic over-farming and drought – which I certainly knew all about – the Big Five were no longer around. The last Cape lion had been shot about a hundred and fifty years ago.

Not only that, commercial farming was turning the land into a dustbowl. Much of the Eastern Cape is not suited for agriculture. It is a magnificent natural wildlife zone with valleys and thickets that bind the soil. Remove that, and the land will wither and die – which was exactly what was happening with agriculture. We farmers were the problem.

My father, brother William and I knew this. We spoke about it at length. We watched with interest what Adrian was doing at Shamwari.

When the Merino wool industry collapsed due to the mass production of synthetic fibre in the early 1990s, it was a bitter financial blow to us. We also suffered heavy losses from stock theft. Transporting the animals to the nearest city of Port Elizabeth was a nightmare, and trucks were regularly attacked by stock thieves. They would kill the animals, stuff the meat into plastic bags, and wait for a taxi to pick them up. We would only discover this when we counted the sheep at Port Elizabeth. Backtracking along the route we would find discarded skins, entrails and limbs. The skull was always taken because in the Eastern Cape a sheep's head is a prized delicacy.

In the light of escalating theft and collapsing markets, we decided to go into game ranching, as Shamwari was doing. This meant we had to close down our sheep and cattle business, which had been in the family for five generations, and start afresh. We also would have to re-wild the land to its former pristine wilderness state, something that had not been done for more than a century and a half.

It was a daunting project, but William, Dad and I had almost religious faith in it. In fact, I believed it was our destiny. First, there was the collapse of my dairy venture, then the Merino wool industry . . . well, perhaps I could be forgiven for believing that fate was channelling me into wildlife conservation.

We had the grand ideas, but not the capital. It would be an expensive business re-wilding 2200 hectares and closing down the family farm at the same time. In other words, we would be spending a lot of money with not much coming in.

Even more problematic was that neither William nor I were living at Leeuwenbosch at the time. I was still farming bananas in KwaZulu-Natal, and newly married William was a struggling veterinarian in Southend-on-Sea on the Thames estuary in England. Neither of us had anywhere near the capital required to do what had to be done.

How to raise the money was the most pressing problem of my life at the time.

One of the answers was to go back to what I knew best.

Goats.

CHAPTER FOUR:

Goats and the Sons of Shaka

'Mayday! Mayday! I'm under attack,' I shouted over the Farm Watch radio.

Gunfire blasting in the background added authority to my call. Outside three or four men were trying to shoot the lock off my office security door. I had been working late and they had arrived under cover of darkness in a pickup truck. Initially I thought little of it when I saw the vehicle stop outside. Then several men got out and a chill ran down my spine.

They were armed. They walked straight up to my office building, fortunately some distance away from the main farmhouse where Angela and the kids were, and started firing at the security door.

Somehow the locks held. I grabbed the Citizens' Band microphone just an arm's reach away and clicked the radio over to the emergency channel linking directly to Farm Watch, a civilian armed-response network that invariably arrives at rural crime scenes long before the police.

As I again barked out a Mayday call, more shots echoed into the night as the gunmen continued trying to blast out the locks. Cordite choked the air and the door would soon cave in.

I pulled out a .9 mm pistol from my desk drawer, but I knew that not only was I outnumbered, I was outgunned.

A cool, unruffled voice came over the radio. 'Copy that, Grant. We're on our way.' You would have thought the guy was giving me a weather report he was so calm.

'Make it quick,' I said, pointing my gun at the door, finger tightening on the trigger, ready to fire if it burst open.

Several minutes later I saw car headlights bouncing up the dirt tracks through the banana fields. The driver had his foot flat on the accelerator. That's one thing about South Africans – when you are in trouble they drop everything to help you.

The gunmen fled. In fact, their departure was so swift that they left their truck behind, sprinting straight into the bush. The Farm Watch first responders, ordinary farmers like me, fired some shots into the darkness to speed them on their way. Then it was quiet.

I was lucky. I probably would not have survived a close-quarter gunfight, if it had come to that. And it made me realise exactly how risky my increasingly lucrative goat business had become.

After arriving in KwaZulu-Natal in 1991, I found that my 'fame' as an Eastern Cape goat farmer had followed me. The province was the ancestral homeland of the Zulu tribe, a Nguni people like the Xhosa with similar language and traditions. This meant that I had a running start in dealing with them from the word go.

The Zulu are even more enthusiastic about goats than the Xhosa. Barely a ceremony goes by without a sacrifice. The animals are always eaten afterwards, so it is part of their staple diet and not solely a blood-letting ritual communicating with ancestral spirits.

I was soon asked by workers on the farms and people in villages and townships if I could get them quality goats. It didn't take a genius to figure out that with my background this was a solid business opportunity. I still had my herd at Leeuwenbosch, and it wasn't long before trucks loaded with goats, driven by an extraordinary man called David Nokanda, were thundering up through Lesotho en route to KwaZulu-Natal a thousand miles away. The shorter road through the Transkei was so riddled with potholes, stray pigs, obstinate donkeys and wayward drivers sometimes high on the local weed, *dagga*, that it wasn't worth the risk.

David was born on the same day and year as my dad so they referred to each other as twin brothers. It was a friendship across the colour-line that lasted throughout David's life. Dad often still talks about him.

In those days there were no satnavs to point drivers in the right direction. Yet David, who had no schooling, would climb into the truck and, no matter where I asked him to go, would find the client and deliver the goats.

Within a few years of our arrival, goats were my biggest money spinner, more lucrative than my main crop of bananas. The cash I was making was going to help swell the Fowlds family coffers to finance the re-wilding of Leeuwenbosch. There was no project closer to my heart.

My goats were without question the best in the province as they were organic before the word became fashionable. They grazed freely on our farm and hadn't been cooped up for weeks in pens. More importantly, they didn't die soon after arriving in KwaZulu-Natal, unlike my competitors importing blue and black goats from Namibia. Namibian goats are used to arid conditions as the country is bone dry. Consequently, as soon as they were offloaded in sub-tropical KwaZulu-Natal, they gorged on the lush green grass. Their stomachs couldn't handle such rich pastures and they often died from diarrhoea.

My animals were hardy Boer goats and in huge demand. I also started a relationship with Indian buyers, as Indians were the second largest population group in the province, and during the Muslim Eid festival I supplied most of the mosques north of Durban. Although Eid only happens once a year, it is a huge market.

However, my key market was Zulu, and soon I could speak the language flawlessly. I regularly ventured far into the tribal heart and learned the traditions of the mighty nation that lived in the beautiful hills and valleys. Just as with the Xhosa, this was the bedrock of a priceless education that I could not have picked up anywhere else. Not only that, it was to become a foundation of trust and empathy in later dealings with communities on conservation projects deep in the outbacks.

Although the rest of my family didn't share my passion for goats, Angela and my three daughters at least fell in love with one of them. His name was Dennis.

He was a monster of his breed with horns as curved as the handlebars of a chopped-down Harley-Davidson hog. He was the family favourite, so when the kids were asked to bring a pet to school for a fun day, they took Dennis along – much to the initial consternation of teachers who were expecting something a tad smaller. He arrived in dignified splendour, standing aloofly apart from the lesser creatures, mainly puppies, kittens and rabbits, that were romping about on the grass. There were even guinea pigs and goldfish.

But he was actually a gentle soul, allowing the kids to ride on his back in return for loaves of bread that he devoured with relish. They loved it, and on future 'pet days', Dennis the non-menace was always summoned.

Dennis was a *kaparter* (castrated male) so he didn't smell foul like a billy goat that tends to urinate on itself. He also was what is called a Judas goat, an animal trained to associate with sheep and cattle and herd them to the abattoir on slaughter days – hence the name Judas. But as I sold our goats live, Dennis didn't do much Judas work, and was only required to lead them to various holding pens.

It was a sad day for the family when Dennis went to the big goat kraal in the sky, and I can still clearly picture his magnificent handlebar-sized horns that a Hells Angel biker would have killed for.

Eventually, my goat business grew so much that I started giving credit, which of course created problems of its own with repayments. I had to enlist a beefy ex-wrestler as an enforcer to sort out bad debts. He wasn't very successful as the townships are such a maze of alleys and pathways that it's extremely difficult to track down someone who doesn't want to be found. However, I was selling more than five hundred goats a week so was able to write off bad debts as business was booming.

Eventually, my sources in the Eastern Cape were depleted and I started importing from Namibia like my competitors, becoming a wholesaler in the process.

This also generated problems, which were somewhat more serious than debt collecting. I was shot at several times as people now knew I was carrying wads of cash in my truck. I was also travelling to some of the wildest places in the country where few white people ventured, so was a visible target. It got so dangerous that I had to hire cash-in-transit vans with bulletproof windows.

Consequently, being attacked in my office by gunmen one night was not a complete surprise. This was nothing personal; they were just following the money.

Even worse was when my right-hand man, Dumisani Mthethwa, was hijacked and beaten to within an inch of his life. His unconscious body was dumped at the aMatikulu river mouth, a lawless chunk of real estate and a favoured area for hiding corpses as they were rarely found. His attackers thought he was dead.

But Dumisani is as tough as rhino hide. Somehow he managed to regain consciousness and limp out of the thick bush to get help.

His attack shook us badly and I installed electronic trackers in our trucks. Despite that, we still had about four or five further hijacking attempts.

The final straw was not gunmen or hijackings. It was bureaucracy, when the government, in its wisdom, decided to impose Value Added Tax on goat sales. Beforehand goats had been zero-rated as they were listed as a basic subsistence food, consumed by the poorest of poor South Africans.

No longer. Overnight, I had to pay 14 per cent on all stock, which slashed my profits to the bone. I was bringing in a thousand goats at a time from Namibia, and now having to pay VAT upfront at the border post was killing me financially. Goats are mainly bought for traditional slaughter, and if you don't sell them within a week of delivery, they lose condition in the pens and the dealer loses money. So to bring in a thousand animals, pay VAT upfront, transport them a thousand miles and then risk losing a significant percentage due to stomach ailments as they were not acclimatised to KwaZulu-Natal grazing, was not the ideal business plan.

On top of that, both Dumisani and I were lucky to have escaped with our lives. The stakes were getting too high and the profit margins too low. It was now time to move on.

Looking back, it had been one hell of a rollercoaster ride. It also in the long run incrementally helped achieve our goal of re-wilding our farmland in the Eastern Cape. Just as goats had paid for my school fees, they were also supplementary to the success of creating our game reserve. Although I must admit, my brother William practising as a vet in the United Kingdom and earning foreign currency – a pound was worth about 12 rand at the turn of the century – was the main contributor. However, he was pining away in the drizzle and cold, treating pampered pets, while I was having far more fun handling goats in the wild sunny outbacks, despite the rocketing bills, flying bullets and bad debts.

Our first tentative steps in converting our farm into a game reserve project were now starting to show promise. With my goat business having run its course, and William about to come home to South Africa, the next chapter – an absolutely key one in the lives of the Fowlds family – was about to kick off.

We had the start-up capital, we had the expertise, we had the land. Except not enough land for our vision of a Big Five reserve. We needed to persuade our neighbours to climb on board.

Now was time to roll the dice.

CHAPTER FIVE:

A is for Amakhala

Anything worthwhile in life involves hard work.

Re-wilding Leeuwenbosch proved exactly that, but eventually we started to see modest results. We brought in a couple of giraffes, twelve zebras and a herd of impala and blesbok and we were in business, although not with many clients. Our biggest money earner came from converting one of the old farm buildings into a boutique guesthouse, where most of our initial customers were overflow guests from Shamwari next door. Shamwari was often operating above capacity and its growing success convinced us we were on the right track.

It also gave us the confidence to approach our neighbours, the Gush, Weeks and Hart families who – like us – came from settler stock that had farmed their lands for several generations, spanning well over a century. If we could persuade them to join their properties with ours and take down fences, we would have the space to create a pristine conservation area and world-class game reserve.

At that stage, we were the only owners with wild game on our land, although it wasn't exactly the Serengeti. A handful of giraffe and antelope was hardly going to lure rich overseas tourists away from the growing number of five-star resorts that were home to the Big Five.

The three families agreed to join us, and wildlife expert David Peddie drew up an agreement whereby we pooled the land of each owner and shared the cost of game and levies on a per hectare basis. It was that straightforward.

However, unlike us, the other families all had a Plan B. They kept their dairy and cattle herds in case this crazy wildlife idea dreamed up by the Fowlds boys crashed and burned. For us, there was no backup option. Re-wilding our land was the only plan. We ploughed every spare cent we had into our vision. There was no going back.

From those humble beginnings, the seeds of Amakhala Game Reserve took root. *Amakhala* is the Xhosa name for the majestic flame-red aloes that grow in the area. For us, it had the added benefit of starting with an 'A', which we thought would be useful for internet searches. But we were just a bunch of guys from the bush – what did we know about fancy digital stuff like search engine optimisation and algorithms?

My dad, William and I wanted to expand further, incorporating more landowners that surrounded the nucleus of four family properties. We soon found one.

It came from an unlikely quarter – a former ace surfer, Paul Naude, originally from Durban but now living in California. Paul, along with the family of his best friend Derrick Cook, had bought the neighbouring farm HillsNek and we discovered they were keen to become part of Amakhala. Paul is well known in the global surfing community and had founded a magazine called *ZigZag* that in the 1970s became a cult among the endless summer surf crowd. He was also an astute businessman, heading up the United States operation of

Billabong beachwear. We were tremendously fortunate to have him and the Cook family join us. Not only did they add a large chunk of wild land to Amakhala, but Paul, Derrick and his son Brent injected a large number of much-needed game into the system.

Despite that impressive boost, the founder families, amicably referred to as 'The Settlers', initially did not want new members. At that stage, range expansion was not considered as important as control of the project.

I thought we were being short-sighted and had for some time wanted to buy the farm Kraaibosch next door to us. It was a vital part of completing the jigsaw of our ultimate vision.

I soon got my chance when I heard that a wildlife expert from Mpumalanga, Will van Duyn, was looking for land to run a disease-free buffalo breeding scheme in the Eastern Cape. We formed a company to buy Kraaibosch, which would bring an extra 1000 hectares of wild land into Amakhala, with the added bonus that it was perfect for buffalo.

This was great in theory, but reality soon rudely interfered. The truth of the matter was that we had no money. The bank flatly refused to lend us a cent.

It looked like our buffalo deal for Kraaibosch was dead in the water. But the gods of good fortune were smiling when Adrian Birrell, scion of a famous Eastern Cape cricketing family, and his wife Sue came to the rescue with the equivalent in those days of close on £18,000 as collateral. Adie is my first cousin and is famous for being Ireland's World Cup cricket coach in 2007. Ireland is certainly not renowned for its bat-and-ball prowess, but under Adie the lowly Emerald Islanders felled mighty Pakistan in one of the game's biggest slingshots ever fired at a Goliath. To this day, he never needs to buy a pint of Guinness in Dublin – which is actually a bit of a waste as he's a teetotaller, and has been ever since I got him drunk in the bush when he was fifteen years old.

Just as Adrian saved Irish cricket that day, Adie and Sue did the same for us. Suddenly the banks were falling over themselves and we did an incredible deal with an extremely awkward seller to buy Kraaibosch at a record price of only £150 a hectare.

Will moved his buffalo to Kraaibosch, giving us the second of our Big Five. We knew we already had leopard as we found both tracks and kills in the bush, although no one had seen one.

Will was a brilliant animal person, but by his own admission he was cantankerous, volatile and not good with people. He was not everyone's favourite, and often rubbed people up the wrong way. But no one denies that he and his wife Sharon played an absolutely vital role in the Big Five development of Amakhala.

We then sold a section of Kraaibosch to close friends and pilots from South African Airways, Captains Neil Wallace and Archie Bell. This had the added bonus of getting us great publicity, as whenever one of them flew over Amakhala, he instructed his passengers to look out of the windows and see what they were missing if they did not visit the reserve. Talk about getting orders from above.

Other members were Andri Schoeman, Deon and Elza Truter, who built the five-star Bukela Game Lodge, and Terry Collinson, a Yorkshire-born man from Guernsey Island. To be blunt, Terry was the sole member swimming against the tide in our organisation. I guess in the law of averages there must always be an opposition party, but while Terry was good fun in a bar environment, he sometimes made running a reserve awkward.

The Kraaibosch splinter group originally fell under the Bushman's River Constitution, an offshoot of the main Amakhala structure, and committee meetings were sometimes as wild as the land outside. In fact, some Bushman's River committee gatherings ended up in shouting matches that would have put the craziest House of Commons histrionics to shame. I regularly had to mediate in very difficult situations, and soon learned that people were far harder to manage than animals. I only discovered later in life how critical a learning curve that proved to be for me. In my later career as a full-time conservationist, mediation became my key focus and one of my main challenges – due to something very aptly called human–wildlife conflict.

Andri, Deon and Will van Duyn later sold their Amakhala properties to the Lion Roars Group, which specialises in hotels and lodges owned by the Bailey family, who originally came to South Africa from Zimbabwe.

This gave us a professional marketing edge as the Baileys are superb hoteliers and lodge operators with a passion for wildlife. As a result, the project is not only a pioneer in re-wilding degraded farmland; Amakhala is today an award-winning Big Five game reserve acknowledged around the world. It's not me saying that; every tourism and hospitality website including TripAdvisor agrees.

However, we certainly paid our dues. The first game drives were conducted by sunburnt farmers in shorts and *veldschoen* (bush shoes) with guests crammed in the back of pickup trucks, or bakkies as we call them. Our neighbour Rod Weeks, patriarch of the Weeks family, still laughs when he says, 'You sometimes had to drive through the cows and sheep to get to the zebras.'

A lot of different factors came into the mix to create Amakhala, and there have been scores of other key role players and visionaries. The story is bigger than any one person. I am merely recounting my intensely personal odyssey of how I got there. In essence, it is the story of my magnificent goats – without them, I would never have earned the funds to buy into the project and I would not be where I am today. Goats will forever be among my most treasured animals.

My dream of getting involved in hands-on conservation was now a reality. It had finally happened.

CHAPTER SIX:

The Big Five

Amakhala was now up and running. We had land, lodges ranging from three to five stars to cater for all categories of wildlife lovers, and most importantly we had animals.

But we didn't have the Big Five. Without that we were stuck in reverse gear as far as international tourism is concerned.

However, we still had one unique selling point. The Eastern Cape is malaria-free. No other Big Five reserves in the traditional game areas of KwaZulu-Natal and Mpumalanga, which includes much of the Kruger Park, can claim that. It is an invaluable asset and something we had to work on.

Our neighbour Shamwari had already made a start on this by promoting their malaria-free reserve at the Africa Travel Indaba, one of the largest marketing events on the tourism calendar. General manager Joe Cloete, who was married to my sister Jayne at the time, cunningly placed a 'no entry' sign with a red line struck through a blow-up of a mosquito on posters at their exhibition. This irritated other reserves intensely, but they could do nothing because it was true. Little did they know that we were going to do the same.

However, fancy lodges, beautiful scenery, glorious African

sunsets and even being malaria-free were still not going to get us the same traction overseas as the top Big Five reserves. We had leopards, which were wild as hell and largely invisible after being shot as vermin for more than a century by farmers, and now the buffalo Will van Duyn had introduced. We desperately needed elephant, rhino and lion. Otherwise there was no global selling point.

The first breakthrough was with elephants. A reserve in KwaZulu-Natal called Phinda had surplus animals and was happy to supply us.

However, the biggest problem was not relocating the animals. And it completely floored me as is it came from one of our most influential settler-owners.

Henry Gush, the patriarch of the Gush family, was adamant that a herd of pachyderms would wreck the countryside beyond repair. He didn't want them, and it seemed nothing would change his mind.

This was a game changer – the gut shot to the project that I dreaded might happen. If Henry convinced other settler families to think the same, Amakhala, as far as I was concerned, was doomed.

'They'll *moer* the bush, *swaer*,' he said to me. *Moer* is an Afrikaans word for 'beat up', but somewhat more forceful than the literal English translation.

We all loved Henry. He was a true Eastern Cape character and also, I'm proud to say, my godfather. He and my dad were first cousins and blood brothers, camping, hunting and fishing as children growing up in the wild. However, he was convinced that as elephants bulldoze trees to get at the carbohydrate-rich roots, they would leave a trail of destruction in their wake. I argued that elephants were historically indigenous to the area, as the world-class Addo Elephant Park several miles west of us attested. In other words, the land had supported pachyderms for millennia. Elephants would probably alter the landscape to some extent, but pulling down trees and shrubs also generates new growth.

It was put to the vote. I held my breath. Fortunately, Henry was the only one against the idea. He was overruled.

I then negotiated with a German TV company to film the relocation, and as a pay-off for the publicity, we got the animals free from Phinda.

When the initial elephants arrived at Amakhala in June 2003, to my astonishment Henry was the first person there to greet them. He was as excited as a kid at Christmas as they were offloaded. If they could have such an effect on a crusty but lovable old guy such as Henry, imagine what they could do to visitors on the reserve?

Until his death, Henry referred to them as 'his' elephants. Far from '*moering*' the bush, they became one of his favourite animals.

Moving elephants is a difficult business, as the animals are notorious escape artists. Lawrence Anthony describes in his book, *The Elephant Whisperer*, how his herd broke out on two occasions and he was told by the authorities that the elephants would be shot on sight if they did so again. Anthony kept a two-week, largely sleepless vigil outside the boma – the livestock enclosure – where his herd was kept, pleading with them not to break out again. They listened.

Shamwari also had a bad experience when their newly relocated herd broke through the electrified fences and rangers had to chase them all the way south to the coast, dart them and bring them back. As some of the area was human settlement, it was a potentially serious situation. As it was, one cow was killed, tossed high into the air like a stick when it stumbled into the fleeing herd's path.

A friend of ours at the Kruger National Park said he knew someone who claimed to have a unique relationship with elephants and was prepared to come down and help settle the animals during the relocation. She was an elephant whisperer, if you will, like Lawrence Anthony. However, Anthony, who sadly is no longer with us, was adamant that he had no 'whispering' superpowers – instead, the elephants whispered to him.

Betty, on the other hand, was clear that she did all the whispering. However, we figured she couldn't do any harm and agreed she could come along, not expecting much. She told us she had put some 'stuff' in the water that would help the elephants relax. William, Dad and I shook our heads, grinning among ourselves. It all seemed a bit of mumbo-jumbo juju.

We spent ten days filming the capture at Forest Lodge in Phinda, and then drove the animals in special crates on flatbed trucks close on a thousand miles to Amakhala.

Whether it was our elephant whisperer's mumbo jumbo or not, our jumbos behaved beautifully. They never remotely looked like they were going to try and break out. They were placid and content from the word go.

After we released them, our elephant whisperer followed them as they wandered around the reserve, which broke one of the key protocols of relocating animals. The golden rule is to release and then cut all contact. But our whisperer ignored that. For the next few days she was with them around the clock, giving us continuous updated situation reports, which was invaluable information.

So who knows? Maybe we just had good-natured elephants. But I couldn't help wondering . . . what was it she put in the water?

We also got two elephants from Addo, but sadly one died in the transport truck. It seems as though it lay down in the crate, then couldn't get up. When that happens, the internal organs are crushed by the elephant's weight.

We buried it on Amakhala and an amazing thing happened. The rest of the herd arrived and dug the carcass up, carrying off the bones. From then on, every time the elephants passed the gravesite, they paid their respects, sniffing the bones, and sometimes moving them to another spot. Even though the original elephants were from a herd almost a thousand miles away, they considered themselves family. I wonder how many humans would honour a skeleton of someone they didn't know.

Later that year we got our first white rhino, also from Phinda. I went to load her up with one of the most experienced wildlife vets in the world, Dave Cooper, who gave her a thorough medical check. We were ecstatic when he told us the animal was pregnant.

The following year the first rhino for more than a century to be born on the land now called Amakhala arrived as a slippery bundle in the dust. He tottered around on shaky legs and we named him Geza, 'the naughty one'. He would one day become one of the internet's most famous rhinos, but that was way into the future.

Geza's entry into our lives brought much delight, but little did we know that as we filmed his first few steps, those days of innocence were numbered. Soon afterwards, no one would be able to video any rhino on Amakhala without strict security measures. Poaching in South Africa would become as rampant as gang murders in south Chicago. The only difference is that gangsters are many and rhinos are few. Today, posting a video of a rhino on YouTube and stating the location is the equivalent of placing a full-page newspaper advertisement that there is horn on your land.

A few years after Geza was born, we introduced the highly endangered black rhino. There are fewer than five thousand left in the world, which meant tightening up our already stringent security even further.

Finally, we got lion, the holy grail for many overseas tourists. We started with a pair, but that was all we wanted as lions breed like the cats that they are. Soon two cubs were born. We already had cheetah, which are not members of the exclusive Big Five club, but without doubt among the most spectacular hunters on the plains. Being the world's fastest animal, a cheetah in full flight makes Olympic sprinters look like arthritic snails. Our cheetahs are also highly prized on the wildlife market as, being on Amakhala, they are aware of the dangers that lions pose. Cheetahs are quickly killed if unfamiliar with their far more powerful feline cousins and don't keep out of their way.

Amakhala was thriving and on track to become a prime destination in the Eastern Cape. Our soaring bed-occupancy numbers and comments on travel websites showed we were doing something right. But it was a hugely expensive undertaking. The costs of re-wilding the land, putting up game fences, electrifying the outer perimeter, running anti-poaching patrols as well as hiring expert wildlife consultants were exorbitant. Almost all profits were ploughed back into the business. Most gratifying to William and myself was that thanks to what we were doing, my parents were still able to live in the ancestral Fowlds home on Leeuwenbosch. In fact, they soon became tourist attractions themselves. My dad, known throughout the area as Uncle Bill, is a raconteur of note, usually with a whisky in his hand, and has guests spellbound with his anecdotes of life in general and the Eastern Cape in particular. Embarrassing at times? You bet. Polite language? No. Boring? Never.

Amakhala could not financially support all of us, but that wasn't the point of the project. It was part of our heritage. In Afrikaans, they have the word 'gees', which literally means spirit, but the English word is lame. 'Gees' is far more elemental. Leeuwenbosch was our 'gees'. But it was not my livelihood.

As a result, I divided my time between marketing Amakhala, where I concentrated on the growing South American market, and managing my banana farm in KwaZulu-Natal.

A tragedy then shook us to the foundations. It almost cost us our children.

CHAPTER SEVEN:

A Miracle of Life

Angela and I are blessed with three daughters.

All are country girls at heart, having grown up on our farm in KwaZulu-Natal, climbing trees and running unfettered under big skies. They are free spirits but their mother always kept them well groomed.

They are extremely close to each other, keeping in regular contact despite often being on different continents. The oldest, Jess, is a cross between a businesswoman and an adventurer. She's got degrees in Marketing, Media and Film and was a brand manager for one of the country's largest clothing companies in her mid-twenties. At the age of twenty-nine, she has sailed across the Atlantic, ridden mountain bikes down the most perilous roads of South America, abseiled off sheer cliff faces around the world and scuba-dived in the Great Blue Hole, a massive sinkhole off the coast of Belize in Central America. In short, she's living life to the full and done loads of crazy stuff that would give most parents – including us – grey hair.

Next is Georgina. She is superb with children, and after qualifying as a psychologist she did au pair care for a year with the Rose family in Sellersville, Pennsylvania. We at first thought she was more introverted than her sisters, but she recently surprised us with a hidden streak of high adventure that we never suspected. She is now also travelling the world, working her passage on a private Russian superyacht. To date she has sailed most of the oceans, visiting North and Central America, Alaska, French Polynesia, New Zealand, the Australian Gold Coast, the Maldives and all the South Pacific and East Asian countries. Her quiet, wise demeanour and deep-thinking mind have earned her much respect and she is an extremely loyal friend to her peers.

The youngest – and tallest – is Alice. She is a typical outgoing Fowlds girl and has friends all over the place. I am convinced she inherited some of her paternal grandfather Bill's genius for remembering people's names, places and directions, guiding her mother around long before Google Maps were invented. She is currently in Vietnam teaching English, and like Georgina, is great with kids.

The fact that all three are still with us is, in my mind, a triumph of the human spirit.

On 29 June 2010, Jess, Georgina, Alice and close friend Stacey Dewey were nearly killed. It was a defining moment in our lives and I still go cold when I think about it.

It happened when the FIFA World Cup came to South Africa and the entire nation joined together in an exuberant festival of football and blaring *vuvuzela* horns. The vibe in Port Elizabeth was at fever pitch as Brazil and the Netherlands were gearing up for the quarter-finals at the weekend.

On that fateful day, William was scheduled to dart rhino, which were due to be relocated, at the Great Fish River Nature Reserve. He asked if my daughters and Stacey would like to go along. They jumped at the chance.

Also with the group were eight American students getting frontline African experience with wildlife vets like William. The group left at 4 a.m. and everything was going according to plan. Alice even posted on Facebook, 'I'm on top of the world,' with a selfie of herself on a cliff above the Great Fish River valley, once the scene of many bloody battles between frontiersmen and Xhosa warriors.

Then William's cellphone bleeped. There was an emergency at Amakhala and he needed to get back right away. Schalk Pretorius, founder of Ulovane training academy and one of the most respected wilderness trail instructors in Africa, was guiding a group of students and had been charged by a black rhino. Schalk had no option but to shoot the animal as it came at them like a missile, wounding it in the head. At the last minute the rhino veered off into the bush. There is little doubt that the ranger's quick action saved several lives that day. However, the rhino was still lurking in the veld with a .375 bullet in its skull, enraged beyond reason and ready to attack anything.

Rangers tracked the wounded beast and a helicopter was sent to Grahamstown to fetch William to dart the animal and remove the bullet.

William asked Jess to drive his Land Rover back to Amakhala and the American students would follow her in another vehicle. With Jess were Georgina, Alice and Stacey.

The Land Rover was a long-wheelbase double cab, and as they were coasting along the back tyre blew out with a crack like a rifle shot. Fortunately, Jess was driving relatively slowly, but even so, the ungainly vehicle skidded out of control. It then spun wildly off the road, rolling several times and pitchpoling into the bush before ploughing into a *donga*, or steep ditch.

Jess broke her shoulder in one of the many flips. Shocked and in agony, she somehow crawled out of the upside-down vehicle and phoned us at Amakhala.

That was the first miracle of the day. We were close by when we got her call and could react instantly. My dad and I jumped into the car and reached the accident scene less than twenty minutes later.

My heart sank as I got out of the car. No one could have survived the wreckage before me. The vehicle was on its roof, mangled and smashed like a crumpled can. Smoke and dust were still wafting in the air.

Alice was unconscious and she, Jess and Stacey – the only one not critically injured – were rushed off in an ambulance. Unknown to me at the time, one of the paramedics examining Alice shook his head and said, 'This one isn't going to make it.'

Georgina was trapped under the Land Rover. If it moved an inch or two, it could collapse and crush her. We had to get her out of there fast.

First, we tried to jack the wrecked vehicle off her, but we couldn't get any traction as the ditch was too steep. By now there were about forty people at the scene and with mounting dread I asked that everyone try and lift the Land Rover, making absolutely sure we didn't shift it sideways or it would fall on Georgina.

It didn't budge. Even with every one of us straining every muscle fibre to the maximum, it was like trying to hoist an elephant.

I slid under the car. What I saw will haunt me for the rest of my days. As the vehicle was upside down, battery acid was leaking directly onto Georgina's leg and she was crying with pain. All the tools needed for rhino darting – chainsaw, drugs, dart guns, ropes – were strewn around her. But the deadliest of all was a bowsaw with jagged razor teeth that had cut her neck under her chin.

I tried to calm her, saying everything was going to be OK. We'll get you out of here, I kept repeating.

I spent the next two hours under the vehicle, holding my daughter's hand as her throat bled and battery acid dripped onto us. In the meantime, a woman arrived, a complete stranger, and started praying for us in Xhosa. We bowed our heads. Before any of us could thank her, she disappeared like an angel into the night.

Eventually an emergency vehicle with hydraulic Jaws of Life tools arrived. We had to cut and rip the Land Rover to pieces to pull Georgina out.

Then there was a second miracle. The ambulance driver ferrying Jess and Alice initially decided to go to Grahamstown as it was the closest hospital, even though it was not equipped to handle someone as critically injured as Alice. If she was still alive when they arrived – which they doubted – she would then be flown to the more sophisticated Port Elizabeth hospital by an air ambulance helicopter. The thinking, albeit unspoken, was that Jess was more likely to survive and would get earlier emergency treatment at Grahamstown.

To his eternal credit, one of the paramedics disagreed. He argued that as Jess and Alice were sisters, they should not be separated. They should both go directly to Port Elizabeth.

If he had not insisted, if Alice had been taken to the smaller Grahamstown hospital, it is unlikely she would have made it. As it was, they got her into Port Elizabeth's state-of-the-art intensive care unit with minutes to spare.

When we arrived at the hospital, Alice was unconscious with

only the heart monitor machine indicating she was alive. Her head was the size of two rugby balls. Georgina was also in a serious condition in intensive care, requiring stitches under her chin. She would soon recover and the battery-acid burns would not leave permanent scars.

While Alice was in a coma, Angela and I virtually lived at the hospital. One of us was always by her side. Most of the family was there in shifts around the clock as well. This would not be the first tragedy with children the Fowlds family had suffered, as my sister Mary-Nan and the four Cypress trees in our garden at Leeuwenbosch attest. We spoke to her, urging her to fight, willing her to hang in there, letting her know we were with her. That she must never give up.

Yet I despaired. She seemed too far gone.

Several days later the third miracle happened. I was talking to her as she lay as still as a ghost in the bed when she suddenly squeezed my hand. It was brief, but unmistakable. I have never known such joy in my life.

We were not out of the woods. But we had a glimmer of hope. At the time, that was all we dared ask.

Slowly she regained consciousness. But she had suffered some brain damage and couldn't speak properly. Nor could she write and had to take six months off school. She went into therapy, tackling her challenges with such steely courage that Angela and I could only watch in absolute pride and awe.

She has now made a complete recovery. In 2017 she got her honours degree in educational support. It was one of the grandest days of my life.

We are blessed with beautiful children who have come through hell. I still cannot come to terms with the fact that I almost lost my three girls. I have flashbacks when I drive past the accident scene, which is not far from Leeuwenbosch.

Yet, Jess says that the experience, dreadful as it was, made her into the person she is today.

She said it taught her to live life to the full.

I think that goes for all of us.

CHAPTER EIGHT:

Geza, the Naughty One

Five months after the accident, we had another tragedy. Poachers invaded Amakhala at night and murdered two white rhinos.

This was the first incident of Big Five game poaching we'd had, and it shook the reserve's partners and staff to the core. It was an evil this slice of paradise had not experienced.

The killers arrived with dart guns and chemicals. These were no backwoods amateurs. It was a highly organised operation run by experienced wildlife criminals who were obviously part of an international gang. The horns were hacked off with machetes – or *pangas* as we call them in Africa – and the thugs left under cover of darkness as silently as they had come.

We first became suspicious on a Tuesday when rangers noticed that two rhino bulls, named Chipembere and Isipho, were missing from their usual hangouts. After an intensive search, the rotting carcass of Isipho, the younger bull, was found on Thursday.

Trackers took to the air in a helicopter, and the bloated, flyblown corpse of Chipembere was discovered the next day.

Both animals were lying in pools of dried blood, their faces crumpled in obscene grimaces. If ever there was a damning indictment of the human race, it is the look you see on the mutilated face of a poached rhino.

Both rhinos were crucial to our breeding programme. *Chipembere* is the Shona word for 'rhinoceros', and we all called him 'Chippy'. *Isipho* means 'gift' in Xhosa, and that's exactly what he was. When his mother arrived at Amakhala from KwaZulu-Natal, none of us knew she was pregnant. The birth of Isipho took us all by surprise, and he was growing into a magnificent specimen and would be a trophy breeding bull.

We erected two simple rudimentary crosses to honour Chippy and Isipho at the top of a cliff on the reserve that we call God's Window, overlooking the Bushman's River. It is a beautiful place and on a clear day the view stretches to the heavens. The tranquillity will never ease the rage seething within us, but it is a fitting memorial for our first victims of the rhino wars and visited by thousands of visitors each year.

Four months later, poachers struck at another reserve, but we were again directly affected. The victim was Geza, the first rhino to be conceived and born on our land. He had been sold as part of a breeding project to Kariega, a Big Five reserve about 30 miles east of us, nestling between the Bushman's and Kariega rivers.

It happened on 11 February 2011, and the first we knew of it was when William – who is Kariega's wildlife vet – got a call from Mike Fuller, the manager of Kariega. William set off, thinking that he would be doing a crime scene post-mortem. Then Mike phoned back with words that stopped my brother in his tracks. 'William, he's still alive.'

At first, William didn't believe it. As a wildlife vet, he knew only too well the appalling wounds poachers inflict when hacking off horns, and doubted any creature could survive such barbarity.

'His horn's gone. His face is a mess. But he's still standing,' said Mike.

William jumped into his Land Rover. Half an hour later he was looking at a bloody apparition that he said will 'remain branded' in his brain.

William suspected the animal might be Geza, but it was unrecognisable. In fact, it was hardly recognisable as a rhino.

'His profile was completely changed by the absence of those iconic horns,' said William. 'More nauseating than that, the skull and soft tissue trauma extended down into the remnants of his face, through the outer layer of bones, to expose the underlying nasal passages.

'Initially he stood on three legs with his mouth on the ground. Then he became aware of me and lifted his head, revealing pieces of loose flesh which hung semi-detached from his deformed and bloodied face. He struggled forward and turned in my direction. His left front leg provided no support and could only be dragged behind him. To compensate for this, he used his mutilated muzzle and nose as a crutch and hobbled towards me. His one eye was injured and clouded over, adding to his horrific appearance.'

Torturously, the dreadfully maimed animal staggered towards William. It was unbearable to watch – the animal's agony and anguish were way beyond the imagination of any living creature.

'Were his efforts to approach me a weakened attempt of aggression towards the source of his suffering?' asked William. 'Or was there a desperate comprehension of finality, a broken spirit crying out to die?'

The rhino was so close that William could see blood

bubbling inside the skull cavities. Every breath added to the animal's agony, the chilled morning air flowing over inflamed tissues and exposed nerves.

William could only stand and stare. Shocked, outraged and disgusted to the depths of his soul, he whispered an apology under his breath.

'I am sorry, boy. I am so, so sorry.'

The stricken animal heard him. He lifted his mangled head, looking directly at William.

'I expected at any moment for his suffering to snap into a full-blown rage, but it never came. I backed away slowly and he kept staggering in my direction, not showing any aggression, just one agonising effort after another. For a moment the thought even crossed my mind that this animal, in an incomprehensible amount of pain, acting completely out of character, could be desperately seeking something, anything, to take away the pain.'

William then had to make an awful decision, both as a conservationist and a vet. He knew the animal could not last much longer. He knew the humane thing to do was to put it out of its indescribable misery as soon as possible.

Yet . . . something told him that what he was witnessing was what a complacent world should be forced to watch. That the rhino wars are far more nasty, vicious and brutal than anyone could ever comprehend, and that they are happening every day of our lives. That darted rhinos do not go quietly into the night when their horns are hacked off with grisly blows from a machete or axe. They live for hours, even days, in screaming torment once the sedative has worn off.

This was the horrible, howling reality of wildlife crime at the front line. Only if we can show the world the raw truth of what is happening on our watch will we stand any chance of winning the war for the planet.

On that bleak bushveld morning, William took a deep breath and metaphorically drew a line in the sand. He decided that this magnificent creature's death must not be in vain. That would be compounding the infinite cruelty of the crime. The animal's suffering must be a scar seared on the collective consciousness of humanity. Then at least the horror would have some value – a bellow of outrage echoing around the world.

Whether people would listen, whether they would be able to accept the truth, was another story.

However, this put William in a serious ethical dilemma. To bring this horror story into the air-conditioned living rooms of the world meant that he would have to film it. The logistics were daunting. He would have to contact a professional cameraman from Port Elizabeth, drive him to the remotest part of the reserve where the savaged rhino was bleeding out, and then record the animal's agony. That would take three or four hours at the least. Was it worth prolonging the rhino's indescribable pain in a possibly forlorn hope that people would listen?

'My mind was telling me that to keep this animal alive was wrong, but somewhere inside I felt certain that the story of this despicable suffering could get to even the most hardened minds. The people driving the demand for this bizarre product, who say they take rhino horn to feel good – surely, they couldn't feel good knowing that animals are suffering to this degree at their hands? If they could, in some way, be made to feel part of the massacre, then perhaps this cruel and senseless killing might stop.'

William walked back to where Mike Fuller was waiting. Mike had been among the first to find the disfigured animal staggering in the veld. The shock had been so great that he had remained at the Land Rover. He could not see it again. He had already been to hell and back.

William asked if the reserve would consider allowing the horror unravelling before them to be filmed.

Mike nodded.

While waiting for the cameraman and consulting with other vets if there was any hope of saving the animal, William got the news he was dreading. The rhino was indeed Geza.

We all knew Geza well. His mother Nomabongo – 'the proud lady' – arrived in 2003, and her presence more than any other animal seemed to transform the whole mood of Amakhala's landscape from farmland into wild land. His father was a rhino from the Phinda reserve that I and Dr Dave Cooper had captured and brought to Amakhala almost a decade before.

Geza was named 'the naughty one' in Xhosa because, even as a tiny calf, he was as cheeky as a monkey and would challenge older rhino, then rush back to the safety of his ever-protective mother. He was so playful that at times we thought he was a bit of a lovable show-off.

Eventually the cameraman arrived and started filming. When he finished, William was at last able to end the nightmare. The most humane way was to administer an overdose of opioid anaesthetic. The method would be the same as the poachers' – with a dart. That would be followed by a heavy-calibre bullet to the brain. Then eternal release.

William loaded the gun.

'A sense of relief mingled with sadness, disgust and shame descended over that small piece of Africa,' he wrote in a report called 'Poached' that was read around the world.

'Will this rhino, whose suffering I prolonged, so that the world could get a visual glimpse of this tragedy, end up as just another statistic in a war that rages on? Or, will this rhino's ordeal touch us in a way that compels us to do something about it?

'What I have witnessed ensures that I will never find peace until the killing stops.'

William's words echoed my sentiments exactly. The fight for the African rhino would soon become emblematic of my life. Like William, I cannot rest while this war is raging.

The video of Geza's harrowing last stand went viral on YouTube, but it takes an extremely strong stomach to watch. It was also screened on South Africa's premier investigative TV series, *Carte Blanche*. Geza became one of the first voices for rhinos on the internet. He became a spokesman for his species, showing in graphic, broadcast-quality technicolour what we are doing to our fellow travellers on this planet.

For us it was more than that. Geza was symbolic of our land. He drew his first breath at Amakhala.

William, through no desire of his own, also received widespread acclaim. I know my brother well. He has no wish to be any sort of celebrity. Especially on the back of such an awful tragedy.

But what he did next was even more incredible. It would make him and an amazing rhino called Thandi household names in wildlife circles.

It was a story that became a legend in the rhino wars.

CHAPTER NINE:

The Last Stand

All Big Five game reserves are at the forefront of the rhino wars, but our neighbour Kariega seemed to be in the crosshairs more than most.

Kariega was where the world first got to see the barbaric suffering of a poached rhino in vivid, real-time video horror as Geza lived out his last agonising hours.

Ironically, it is also the scene of one of the most incredible wildlife veterinary triumphs in South Africa, if not the world. It involved a team in the bush led by my brother William, specialists at Onderstepoort Faculty of Veterinary Science, a laboratory in Ohio, USA, and a Cape Town plastic surgeon.

In March 2012, almost a year after Geza was murdered, poachers darted another three rhinos at Kariega. They then hacked their horns off with machetes, somehow evading around-the-clock security and armed patrols.

One male died that night. But when William was called out the next day, the other two animals, a male and a female, were still breathing.

Both were critical. The female, whose name was Thandi, meaning 'love' in Xhosa, looked the worst with a bloody, pulped mess of a face. The male, Themba, which means 'trust', looked marginally better as his face had not been as mutilated. Ominously, however, he was dragging his left back leg. This often happens to a poached rhino as when the killers dart it, the animal falls awkwardly and circulation to the back limbs is cut off by their crushing bodyweight. Gangrene soon sets in.

William was determined to use every shred of knowledge gained from the Geza tragedy to treat the two battered and maimed creatures lying in the veld. He had brought a professional camera team along to document the event, as survivors discovered in time are so rare that film footage is invaluable for future reference. Also, as Geza had shown, a brutal on-the-spot videotape of an animal's terrible suffering in the bush galvanises a lethargic world far more than any impassioned speeches in air-conditioned halls.

He wanted his team to be as prepared as possible for the gruesome task, but that is an oxymoron. Nothing can prepare anyone for such evil. Everyone, camera crew included, was sick to their souls.

Themba was semi-conscious and initially the team thought he stood the better chance of survival, even though he could not stand properly. Thandi's wounds made a Halloween mask look like a Barbie doll. Raw flesh was hanging off her face in loose, bloody chunks of meat. William studied her for a moment.

This one we will definitely have to put to sleep, he thought.

The prospects looked grim. However, her legs seemed fine. She had not fallen at an angle like Themba when darted. Perhaps that was a glimmer of hope. If so, it was the only one.

The next few hours were crucial. William and his highly motivated team decided to give the rhinos a reversal drug to wake them fully and see what could be done. The drugs, both sedation and reversal, are extremely potent and only available to veterinary surgeons. The fact that the poachers used the same opiate as vets was a stark indication of the extent of corruption prevalent on the black market.

To William's astonishment, both animals regained full consciousness quickly. Within five minutes they were struggling to stand, which was too soon as William wanted to do more tests.

For anyone with a shred of conscience, watching the pain kick in as a mutilated rhino comes out of a drugged state is distressing in the extreme. William now had to make the same dreadful choice that he had faced with Geza. Should he prolong the agony and try and save the two animals? Or should he humanely end the creatures' misery right away?

With Geza, he had reluctantly extended the mangled rhino's life in an attempt to expose the cruel reality of the wildlife wars. He could not save Geza, but the tragedy had touched a massive number of people. Geza's horrible death at least had some meaning, a raging scream rebuking a world that allowed this to happen.

William considered the two options, both equally awful. Such decisions torment him, as they would any vet. But as they had reached Thandi and Themba less than twelve hours after the attack – much sooner than with Geza – maybe, just maybe, this could have a different outcome. No one knew.

William decided to make no immediate decisions. He would see how events played out, minute by ghastly minute, hour by harrowing hour. If Thandi and Themba showed that they wanted to live, he and his team would try to honour their wish.

Thandi and Themba did just that. They showed in spades how precious life was to them. The two young rhinos fought with every fibre of their being. Their courage and granite determination should be an inspiration to the human race, that life is a gift to every sentient creature.

Instead, the horns of their species end up in some alleyway shop in the Far East. Sold as a supposed aphrodisiac or hangover cure, or something equally spurious.

For William and his team, top priority was to manage infection and they sedated the animals again to cut off the rotting layers of dead flesh seething with maggots. Once the necrosis was removed, they could start regenerating growth.

'I knew I could not do it myself,' said William. 'So I brought in two top veterinary surgeons from Onderstepoort, Johan Marais and Gerhard Steenkamp, as well as a specialist in human plastic surgery. A laboratory in America scanned in a replica of a rhino skull with all the organs so we could digitally see how we had to restructure the face.'

The first challenge was to close the gaping hole in the hacked skull where Thandi's horn had been. The wound was 12 inches long and 6 inches wide, exposing the underlying sinuses. Infection was spreading rapidly and they had to dig out the clusters of maggots laying eggs in the machete cuts.

'We sliced away as much of the dead tissue as we could,' William wrote in a Facebook blog that was getting masses of responses a day. 'Her fighting spirit is humbling to witness. The whole team is deeply moved by the horrendous injuries and the bravery of this soul.'

William also decided to let the animals heal in the wild rather than move them to a boma clinic, which would have made it more convenient for the veterinary team. He knew from past experience that wild animals get massively stressed in captivity, and any further strain on these two valiant creatures would surely kill them, finishing the job the poachers had started. In the wild they could also retain social structures with other rhinos and be in familiar surroundings.

Once the wounds were cleaned, Thandi's sinus gaps started closing over. Remarkably, she started to get stronger.

Sadly, the same was not true for Themba. He still had little blood circulation to his leg, despite managing to limp several miles on some days. His determination was absolute, but William and his team reluctantly started to reconsider euthanasia.

However, after each procedure when they cleaned the young bull's wounds and got rid of maggots, he always bounced back with such gusto that the team dared to hope he might make it after all.

That was the biggest factor influencing William. Themba still had a clear and unyielding will to live. There was no indication that he was giving up. He was drinking from a nearby waterhole, grazing and moving as much as he could, despite the fact that he was lame.

The audience of William's regularly updated blog was by now expanding exponentially; it was avidly read by animal lovers around the world, daily logging onto computers and cellphones to follow the progress of two mutilated young rhinos in the African bush. On 6 March 2012, he wrote, 'At 2 a.m. my thoughts could be condensed in one line. A small flame of hope draws us forwards into another shameful day of suffering at the hands of humanity.'

Those poignant words touched the hearts of many. Thousands of people of different nationalities were rooting for Thandi and Themba thousands of miles away. And watching the animals fighting for their lives under the blazing late summer sun in the Eastern Cape savannah, it seemed the two rhinos knew it.

Later that day, William and Kariega manager Mike Fuller cautiously approached the young bull rhino.

'Mike and I crouched seven yards from Themba in a small opening in the valley thicket. We watched him for more than ten minutes. I couldn't help feeling that this animal understood something about what we were trying to do for him.

'There is a side to a rhino that is as soft as the mewing call they make. I used to think that sound didn't match the animal, but as time passes I am starting to think that it's more than likely spot on.'

Sadly, the two men noted that Themba's leg was now completely starved of blood. Every step must have been unbearable pain, but still he continued fighting for life, limping to drink or graze.

The decision to play God or not was taken out of William's hands. Several days later they found Themba dead in a watering hole. His leg had finally collapsed. Unable to rise, he drowned.

There was silence at the campsite that night. Some members of the team were in tears. They knew they had witnessed something extraordinary; a display of unimaginable courage and dignity. There had been nobility amid such bleakness.

Thandi was still alive. The small flame that William wrote about in his blog still flickered. For the men and women helping him, not to mention the thousands of followers of his blog, it was little short of a miracle.

William wrote that night, 'Having had my fingers buried in her facial wounds down to my knuckles, I find it hard to believe that she can be still walking around, alert and responsive to her surroundings. The fact that she appears to be doing so well should mean that during the next two or three days we can start on the next phases of her treatment, which is going to be traumatic in itself. She has a shelf of bone slashed by machetes, half an inch thick and four inches wide, still partially attached by connective tissue that I don't think will survive and may have to be removed by a saw down to viable bone.

'Then dead and dying cells have to be scrubbed off under anaesthetic so that a clean bed of healing tissue can get established. This will have to be repeated several times. As bad as this sounds, nothing she still has to get through can be compared to what she has already conquered.'

Thandi grew stronger by the day. It seemed that she would live. But she was not out of the woods. The team still had to assess the long-term effects.

'We have shown we can repair a face,' said William. 'But Thandi still has to be able to withstand the rigours of rhino life.'

Some months later, when Thandi appeared to have fully recovered, she was introduced to a bull. It was a mistake, as rhinos are tactile creatures and her face was again injured by normal male and female interaction.

William and his team then re-grafted more skin and replanted some of the back plating of her horn to strengthen the front section. They also sanded down the stump of the male rhino's horn to have less impact on Thandi's face.

This was a crucial test. If veterinary surgeons could not fix this, the profession would possibly have to re-evaluate the benefits of saving poached rhino. There was little point in rescuing a highly social creature if it could not interact with the rest of the herd.

But the re-grafting surgery worked in the most beautiful way possible. In 2013, William took blood tests that showed Thandi was in the early stages of pregnancy. This was a marvel in itself.

However, rhinos gestate for sixteen months so it was going to be a long wait.

On 13 January 2015, Thandi gave birth to a healthy female rhino calf. The baby was named Thembi, meaning 'hope' in Xhosa.

For William and his team, the birth was the culmination of a long hard road. It had put the rhino wars on the world map with an intensely personal tale of two extremely brave animals, and made our patch of the Eastern Cape world-famous. William was invited to speak at international seminars. Two years later, Prince Harry visited Amakhala, spending two nights being regaled by my father with wild stories of Africa that he had never heard before and probably never will again. David L. Robbins, an international bestselling author, came out from Virginia in America to consult with William and get a first-hand, vivid and authentic understanding of wildlife crime for a novel, *The Devil's Horn*.

But William would far rather be doing what bush vets are meant to do: initiating rhino breeding schemes, or introducing the highly endangered species to new habitat. Instead, he is being called out to investigate poaching crime scenes, do media work, and treat the most gruesome wounds imaginable. Sadly, that's now part of a modern wildlife vet's job.

'Thandi is a living testimony of what our species is doing to hers,' he said. 'She is a reminder of what depths humanity has sunk to on our planet. She represents the real story of rhino suffering.

'The terrible thing for me is the deep sense of shame. Rhinos do not know who is a vet or who is a poacher. All they know is that we smell the same. We are the same species, and the sense of shame is awful to bear.'

I was farming in KwaZulu-Natal at the time and followed Thandi's and Themba's progress closely. What William and his team did at Kariega was, in my opinion, beyond belief.

Little did I know that later that year I too would be sucked directly into the rhino wars, in a way I would never have imagined.

CHAPTER TEN:

The End of the Farm

While William was performing miracles in the bush, Angela and I got news that would irrevocably change our lives.

Our banana farm, Doringkop, our home for almost twenty years, was to be expropriated and given to people with a historic claim.

Land is an emotive and hugely complex issue in South Africa. Suffice to say that the fact you have a title deed and legal bill of sale is not always a guarantee of ownership. There was nothing I could do about it. I was offered the going rate, as opposed to the market rate, and perhaps if I had not been so involved with marketing Amakhala and pursuing my conservation plans, I might have hung on for a higher price. But the bottom line was that we had to leave.

Ironically, I would have many dealings with land expropriation and tribal claims in the future, but thankfully in a far more positive sense in that the new owners wanted to build game reserves.

Part of the deal was that I was to mentor the future owners of my farm for the next two years on the intricacies of banana growing and land management, so it was not as though things were going to happen overnight. However, I had to start thinking of alternative career options.

These were uncertain times. Then, to cap it all, in September 2012 burglars broke into our home and ransacked the property in a particularly nasty home invasion.

Fortunately, I was away in Mendoza, Argentina, on a marketing trip for Amakhala. On that visit I was a guest of Tony Leon, the South African Ambassador in Argentina and former Democratic Party leader who had become a good friend. Earlier we had watched the Springboks play the Pumas, the Argentinian national rugby side, with the snow-capped Andes mountains towering in the background. Little did I know that while I was being wined and dined thugs were pillaging my house across the Atlantic Ocean.

Infinitely more fortuitously, Angela and the children were not at home. My family never stayed in the farmhouse if I was away and were with Angela's mother when the robbers attacked. Thank God for that.

The thugs smashed the steel security barrier with an axe, then chopped the back door virtually off its hinges. From there they hacked through the roof ceiling to smash the burglar alarm – which possibly indicated they had been told about our electronic security system.

Then they went on the rampage. They were looking for cash and valuables and ripped, tore or wrecked everything else. They crowbarred the safe off the wall, smashed it open, threw bedding all over the place and even turned the piano upside down. All our possessions were trashed.

What was most surprising was that we had a large Rottweiler outside in the yard, and for some reason he didn't raise the alarm. He hadn't been tied up or injured by the thugs. Yet he didn't put up a fight, which was totally out of character. Again I wondered . . . was it an inside job? Who knows?

A home invasion is the ultimate assault on personal privacy, and Angela was understandably shocked to her core. Most sinister of all was that her passport had been thrown onto the wrecked bed with an axe placed on top of it. The symbolism, intentional or not, was stark.

Police caught the crooks, and some of the guns, fishing rods, binoculars and other stuff were recovered. I still feel shaky when I think of what could have happened if Angela and our daughters had been in the house at the time.

In the past, we had always felt safe there, even though it was in the middle of a banana plantation. I knew I was respected among the locals as I spoke their language, always stopping to chat about their families or the weather, and formerly did direct business with them selling goats. I regularly came home late and in pitch darkness after playing squash at a nearby country club without any concerns. Our security gate didn't even work.

No longer. That illusion was shattered for ever. Angela doesn't scare easily, but as far as she was concerned, this was it. After seeing the trashing of our personal belongings, she packed everything that was still salvageable, locked the doors and waited for me to get back from South America.

She never went back to the farmhouse again.

I continued living at the house as I was still mentoring the new owners, but Angela and the girls moved in with her mother Eunice on the KwaZulu-Natal Dolphin Coast just south of Ballito.

Six months later, in February 2013, my mentorship came to

an end and I moved off what had once been my dream farm for good.

That was the end of my agricultural career. But what was I to do next?

For that I have to thank an inspirational man, a force of nature called Kingsley Holgate.

CHAPTER ELEVEN:

Rhino Art – Drawing
for Species Survival

Kingsley Holgate is a giant of a man, standing more than 2 metres tall if he doesn't comb his hair, which is not infrequent.

His impressive presence is further enhanced by a white spade-shaped beard smothering his tanned chest. Silver hair, as luxuriant as a rock star's, tumbles down to his shoulder blades. He could be mistaken for a poet, a pilgrim, a swashbuckling adventurer, an explorer or a monk.

He is a mix of all of the above. His poetry is in his storytelling. Kingsley in full cry, holding a Zulu *assegai* – a spear – while recounting epic voyages around a fireside under Africa's megawatt stars, is mesmerisingly eloquent. He is like a monk with his spiritual belief in the soul and beauty of Africa. He is an adventurer in the truest sense; doing things because they are there.

But he is best known as an explorer, often dubbed Africa's most travelled man, and is a Fellow of the Royal Geographical Society. His explorations, many of them world firsts, are more personal odysseys than expeditions and are the stuff of legend. He's rubbed shoulders with Somali pirates, child soldiers, wildlife poachers, villains and rogues. He's broken bread with unsung saints and unassuming community heroes who would be media celebrities anywhere else.

In short, he's met more interesting and diverse characters – like himself – than possibly anyone on this continent.

Kingsley believes in improving lives through adventure, so his expeditions are about philanthropy as well as adrenaline. He provides glasses to rural people with bad vision in a 'Right to Sight' project. He donates mosquito nets to communities ravaged by malaria. He distributes purifiers to rural people whose water is contaminated.

He is also an eco-warrior, and his project Rhino Art has arguably done more to highlight the plight of Africa's most endangered animal among schoolchildren than anything else.

The first person I phoned when my banana farm was expropriated was Kingsley. I had got to know him more than twenty years earlier, when Angela and I joined his supper club, a lively – or perhaps more accurately, eccentric – bunch of people who take it in turns to serve dinner at each other's homes. These are memorable as well as gastronomic events, since with Kingsley the biggest crime anyone can commit is to be bored.

As we both love Africa, its people and wildlife, Kingsley was a good sounding board for ideas on what I should consider doing next. Angela and his late wife Gill, whom we called *Mashozi* (the Zulu word for 'she who wears shorts'), had also become good friends, so we were close as families and I knew I would get good advice.

However, I was taken by surprise by his answer. Instead of providing a list of options, he suggested I come and work with him. He is a partner in the Shakaland Hotel and Zulu Cultural Village complex and wanted me to take on a role as a part-time project marketer. As I spoke the language fluently and had an intimate understanding of Zulu culture, Kingsley believed I would be a good fit in their setup.

Kingsley started Shakaland some thirty years ago with his good friend and fellow hellraiser Barry Leitch. Barry is as close as anyone will ever get to being a white Zulu. He grew up on a farm near Melmoth, not far from the historical Zulu capital of Ulundi. But Barry was no mere academic observer of Zulu ways; he was an active participant, learning the intimidating foot-stamping *impi* – or warrior – dances and competing in stick-fighting competitions. These are not for the fainthearted. Sticks are ferocious weapons in Zulu hands, and blood is always spilled. Injuries are sometimes serious. Barry was the champion fighter in the area, which carries huge prestige and something I don't think any white has achieved before.

Kingsley, who has known Barry for most of his life, was best man at Barry's first wedding with strict instructions to dress properly and behave accordingly. Barry's mother apparently was wary of Kingsley as he was always up to mischief, perhaps mistaking his exuberance for bad behaviour.

Kingsley arrived, dressed in top hat and tails as per instructions. Only problem was that underneath the fancy coat he wore a torn yellow King Corn T-shirt, advertising a popular brand of the staple Zulu diet. He had a beer in his hand. It was not the first.

It was a hell of a wedding, which goes without saying. When Barry and Kingsley get together, even a funeral would end up loud enough to wake the dead.

Shakaland is world-famous as a destination for people

wishing to experience the lifestyle, social system and rich culture of the Zulu people. The kraal where it is situated was used for the filming of the mini-series *Shaka Zulu*, which was a big hit in Europe and America in the 1980s and is still being screened on Netflix.

I accepted the position, and going into partnership with Kingsley was an almost vertical learning curve. He was much more of a mentor than a boss. His passion is second to none. He has an amazing leadership style and never raises his voice, no matter how angry he is. When he said something like 'It would help if you did this . . .' or 'The reason for my call . . .', we all knew it was an order, not a request, no matter how courteously delivered.

From there, events snowballed so fast that I barely caught my breath. Rhino Art starting taking off like a rocket with the increasing number of schools to visit, and Kingsley decided I was the right guy to handle it. I still did work marketing Shakaland – and, of course, Amakhala – but now my main focus was on youth conservation work through Rhino Art.

I was delighted. I knew the project was visionary and, more importantly, it would actually make a difference – targeting a previously neglected audience that in my opinion was the most important one. To get rural youth involved would have far-reaching consequences for the future of Africa's wildlife.

With Amakhala and Rhino Art, I was now completely immersed in conservation. I was doing what I felt I had been destined to do. There was no going back.

Rhino Art started in 2013 on one of Kingsley's expeditions called '*Izintaba Zobombo*' ('The Lebombo Mountains') – a voyage along the spine of the long, rugged range.

The plan was to explore the range, which runs northwards from Hluhluwe in KwaZulu-Natal through to the southern tip of Zimbabwe. Much of the route skirted the Kruger National Park, where game fences that once marked the border of the reserve that is larger than Belgium have now been taken down to create the Great Limpopo Transfrontier Park. This gives the animals far more access to their traditional migration routes, which was great in theory. But the reality was somewhat different. Any rhino that crossed from Kruger into Mozambique had, at best, forty-eight hours to live. For many communities living on the border, poaching was their main livelihood. At the time, a rhino was killed every eight hours with poachers paid a pittance compared to dealers in the Far East who were banking millions.

It was this wholesale slaughter that convinced Kingsley and his team to change the purpose of the *Izintaba Zobombo* expedition. It would no longer be primarily an adventure into a largely unexplored mountain region; it would instead become a conservation project. But it had to be different from any other initiative.

If you want to do something differently, Kingsley is your man. In a stroke of genius, he and his people came up with an idea like no other. He decided to bypass official reports from biologists and rangers and instead get the candid perspectives of the children who lived in this wildlife war zone. In Kingsley's mind, the issues were simple. The youth of Africa were the future – so what did they think about a species that has survived for millennia expiring in their lifetime?

There was only one way to find out. Ask them. Consequently, wherever the expedition stopped, schoolchildren from surrounding communities got a piece of A3 paper and crayons with a request to draw what they felt about poaching.

The results were astonishing. Through their art, sometimes just crude scratchings and scribblings, the children showed staggering eloquence. Many pictures also had a cruel beauty, indicating the depth of talent. It was as though a spigot of creativity mixed with anguish and hope had been opened.

Some drawings showed a rhino with an impressive horn and a man with an AK-47, complete with banana-shaped magazine, riddling it with bullets. In others, there were stick-like figures shot by rangers, or rangers shot by poachers, with waxy-red 'blood' from the cheap crayons dominating the page. But there were recurring threads in all the drawings; animals being slaughtered, helicopters chasing men with assault rifles, stolen vehicles bashing through the bush, and gangster-style gunmen wearing bizarre bling jewellery living in big Portuguese-era colonial houses. The wealthiest poacher in the Massingir area, the key poaching hotspot, was referred to by the kids as Mr T. That information was passed onto the police and even brought up in the Mozambique parliament when Kingsley later met President Joaquim Chissano.

For the children whose homes were in the inferno of a wildlife battlefield, the issues were crystal clear. They had all lost fathers, brothers, friends and acquaintances. They saw this as a conflict between people and environment. The carnage was not only wiping out the animals, it was corroding the core of community life. Mozambique is the only country to have lost all its rhino twice. The animals were completely exterminated during the RENAMO–FRELIMO civil war from 1977 to 1992, and once peace was signed, rhino were reintroduced to their traditional ranges. They have now been shot out for the second time.

This was the first time children had been asked for their views on the rhino wars. The results were so graphic that

Kingsley and his team decided to send the drawings to schools in Vietnam, the main market for illicit rhino horn. The logic was that if Vietnamese children saw what children in Africa thought about the mass slaughter of rhino, they might convince their parents to stop buying horn.

It was ambitious in the extreme but, at the very least, it provided the children at the front line with a platform.

That was the origin of Rhino Art. The mission statement was 'Let the children's voices be heard'. It was a direct insight into the hearts and minds of the children of Africa.

This is where I got involved. Initially Rhino Art was aimed at communities close to game reserves, but I was tasked to take the project to the rest of South Africa, and eventually throughout Africa. Initially I worked with and soon took over from two dedicated wildlife warriors, Carla Geyser and Bronwyn Laing, who gave huge amounts of their time to launch Rhino Art in the formative years.

With me was my right-hand man Richard Mabanga, an incredible person whose enthusiasm for the project is unquenchable. Usually dressed in colourful traditional Zulu attire, Richard is Rhino Art's educator and cultural ambassador, coaching pupils and teachers about the historical and social significance of rhinos to Zulu society. His nickname is 'Mahlembehlembe' – 'all over the place' – and it is apt as he is a ball of energy. He is a gifted, natural performer and feeds off an audience who respond with instinctive enthusiasm. He now speaks fluent French to reach an even wider market.

Richard previously worked for Kingsley as a tour guide at Shakaland, but found his true calling at Rhino Art with his extensive knowledge of Zulu history and culture. When Kingsley asked him to join the team, he knew little about the rhino crisis, but learned fast and is now one of our most effective protagonists against wildlife crime

The kids love him, and wherever he goes they wave until their arms are about to fall off, shouting '*Siyabathanda oBhejane*' (We love rhino).

Richard and I also combine Rhino Art education days with football tournaments, where we provide prizes, balls, referees and even the kit, thanks to the sports company Adidas.

A typical football festival starts with a Rhino Art competition and judging the children's drawings. The winners always get awards, sometimes a bicycle, which is a massive prize for rural kids.

Then the teams are fitted out with Adidas football boots and jerseys, and the games begin. On a good day we would be only three hours behind schedule, due to lengthy impassioned speeches by tribal elders, much dancing and more refreshments combined with the sheer energy of the occasion.

The pitches are little more than a scraped section of bush with rudimentary goalposts. Whoever coined the cliché 'level playing field' has never been to a Rhino Art tournament. Most pitches slope so steeply that it is crucial to win the toss and get the downhill advantage while still fresh.

The prize is sometimes an impala ram donated by a nearby game reserve, usually un-gutted, so the carcass, which would provide enough meat for the entire winning team's families, sometimes starts bloating in the heat.

The games are played with the same intensity as an international, and the losing teams would invariably be in tears. The man-of-the-match award is tantamount to winning the lottery in local status.

Throughout the day, the message goes out: save our rhinos.

This is exhausting work, as it has to be done personally, and Richard, myself and Kingsley attend as many of the art and football competitions as we can. This is not something we can do on a computer. It is not electronic. Every child has to be physically handed a piece of paper with the outline of a rhino and crayons. Every man-of-the-match certificate has to be handed out by someone from Rhino Art.

So far we have reached hundreds of thousands of school-children. My goal is a million.

Then we came up with something even more ambitious: a global summit.

The idea was hatched during a conversation between Kingsley, Sheelagh Antrobus and Colleen Fletcher, the head of the American International School in Mozambique whom Kingsley had met on his expedition across the Lebombo Mountains. She had been with him when he urged President Joaquim Chissano to show more political will in eradicating the poaching scourge in his country. This resulted in a Mozambique Wildlife Summit in Maputo, the country's capital, in 2013. At that event Colleen's pupils plastered the conference arena with their Rhino Art drawings and played a pivotal role speaking directly and powerfully to foreign diplomats, business leaders and Mozambique government ministers.

Kingsley, Sheelagh and Colleen were so moved by the pupils' passion that they had a 'lightbulb' moment to go even bigger: a global summit.

Returning to South Africa, Sheelagh took the idea to the Project Rhino community and the first ever World Youth Rhino Summit was born.

Sheelagh is one of the driving forces in rhino conservation circles and later became Kingsley's partner after his wife Gill tragically died, way too young, after an illness. Sheelagh not only has an innate enthusiasm for everything she does, she's an organisational genius. She had initially been coordinator and founder member of Project Rhino, an umbrella body for various organisations fighting for the survival of the species, before joining the Kingsley Holgate Foundation. On one occasion she raised half a million rand for Project Rhino with a highly publicised skydive. She was invaluable.

I was included in the summit project as I was handling Rhino Art projects in South Africa, but had no idea what I was letting myself in for.

The plan was simple. Under Kingsley's leadership, we were going to invite young delegates from around the planet to protest against the increasing probability that the first large mammal to become extinct since the woolly mammoth would happen on our watch.

This would be one of the biggest ever international wildlife platforms for the voices of the next generation – not only to be heard, but to be heard shouting out at full volume.

The message would be unequivocal: save our rhino.

CHAPTER TWELVE:

Voices of the Future

One thing I learned from Kingsley while doing Rhino Art shows was that it has to be theatre of the highest order.

From the moment we arrived at the school to the moment we left, we had to give it everything we had. We had to hit the ball right out of the park. We were on stage with dancers and singers and storytellers, and it was essential to involve the entire audience, get them singing and clapping and just whooping it up, despite the seriousness of the message. That's the way to reach the children, because when we came again, we wanted to be as sure as hell that they remembered us.

I was soaking up experience from the master. Kingsley is one of the most gifted storytellers in Africa. The *assegai*, a Zulu spear, is his trademark and, when he speaks, he uses it as a prop like an Oscar-winning actor.

I needed to get my own trademark, so one day I arrived at a function wearing Zulu trousers called *mblaselo*. These are jeans or cargo trousers with stitched patches of cloth, beads, bits of animal skin – you name it – creating a surreal fashion style that is uniquely African.

Mblaselo pants have a rich tradition in Zulu culture. It is believed the concept originated with rural men migrating from their huts in the far-flung valleys and hills of their ancestral lands to the teeming First World city of Johannesburg, looking for work in the mines. For them, Johannesburg was known as *eGoli*, the place of gold.

They had no money to catch trains or buses, so they walked. For most, this was a journey of at least 600 miles, and when their clothes got tattered and torn from cutting through bush and thorns or being on the road, they would patch them with whatever material they could get their hands on. These were usually bright-coloured bits of cloth, and a man's *mblaselo* became a work of art – a garment of many colours, like the biblical Joseph.

They were always highly personalised. For example, if a man was a member of a gumboot dancing team, he would stitch a particular chunk of hair from an angora goat into the clothing to prove that. Beadwork from his home kraal was also woven into the patchwork. Most *mblaselo* garments have huge symbolic, almost spiritual, significance for the wearer.

I was no different. The symbolism for me was the African rhino, and not only did I wear my *mblaselo* with pride and respect for the Zulu nation, I sewed on my own patchwork. They became my trademark – my brand, if you will – at all Rhino Art functions.

This was not just in Africa. I would wear them when I travelled overseas on fundraising tours, which always provoked response. People would yell and wave at me in the streets of America and Europe and I would shout back in Zulu or Xhosa. The crowds loved it, particularly people from Africa living overseas.

The only place I didn't get any reaction was on the New York subway. Most people there are way weirder than my trousers, and no matter how crazy my *mblaselo* looked, I hardly stood out among all the purple rooster-comb hairstyles and nose rings. They probably thought I was one of them.

My wife Angela was also not a big fan, particularly when I wore them to the local shopping mall and she would sternly instruct me to wear 'proper clothes'.

The World Youth Rhino Summit had to be done with the same exuberant passion and stagecraft that we brought to our Rhino Art shows. We needed something different. Kingsley's trademark *assegai* and my over-the-top *mblaselo* were good places to start with out-of-the-box thinking.

The organising committee was headed by Kingsley, myself and Sheelagh, and included a broad spectrum of passionate conservationists from Project Rhino and Ezemvelo KZN Wildlife, the KwaZulu-Natal provincial government's conservation authority.

Kingsley and I handled the event's delegations and logistics. Attendance would be by invitation only, and at least 30 per cent of youth ambassadors had to come from disadvantaged communities connected to Rhino Art projects.

We didn't know it then, but we eventually would have to turn hundreds of people away. There simply wasn't enough room.

All this would not be possible without funds and Sheelagh had the most complex task by far in managing the finances. We would have to raise millions, and she immediately started contacting businesspeople and diplomats sympathetic to conservation in general and rhinos in particular. One of her biggest coups was getting the United States Embassy in Durban on board, thanks largely to Colleen Fletcher. As mentioned, Colleen was part of the initial 'lightbulb' moment to host the world rhino summit, and was invaluable as her

American International School provided funds to bring most of the youthful delegates to South Africa.

I was also international coordinator as I had contacts from marketing Amakhala in Chile, Brazil, Argentina, the USA, the UK and Germany. We had a $3 million budget which Sheelagh raised in ninety days, and another $1.5 million that we got from selling rhino-related merchandise.

The next decision was the venue. That was easy, particularly as we had the full support of Ezemvelo KZN Wildlife. We chose the Hluhluwe–iMfolozi Park, as it was there that the white rhino was brought back from the brink of extinction last century by a fiercely committed man, Dr Ian Player.

When Player, the elder brother of world champion golfer Gary, arrived as a young ranger at the game reserve in 1953, there were only 600 white rhino left in the world – all of them in the Hluhluwe–iMfolozi Park.

He launched what was called 'Operation Rhino', one of the most successful conservation operations in history. Player and his team initiated a ground-breaking capture and translocation programme taking pockets of this tiny group of critically endangered animals to other reserves and breeding them. At the turn of the century, forty-seven years on, there were 17,000 white rhinos in nine African countries and nearly 800 in zoos and safari parks around the world, all thanks to Player and Operation Rhino.

The success, unrivalled anywhere ever, has tragically now started to unravel with the current rhino wars. In his final years, Player watched his life's work disappearing before his eyes when the new rhino slaughter began in 2007–8. Only this time it was happening with the aid of First World technology; helicopters, sophisticated tracking devices and assault rifles.

Along with world-renowned conservationist Paula Kahumbu of the Kenyan-based WildlifeDirect and my brother William, Dr Player would be one of the keynote speakers. We did not know when we asked him that it would be the last time he would give a speech.

The summit sparked global interest and we were stunned at the reaction we were getting. We had some amazing personalities getting on board with endorsements from Archbishop Desmond Tutu and Dr Jane Goodall, probably the world's best known living conservationist.

Of specific interest to me was a dynamo called Trang Nguyen, the founder and executive director of WildAct Vietnam, who specialises in exposing the trafficking of endangered species, particularly rhinos, bears, pangolins and elephants.

She had been instrumental in getting the Rhino Art drawings sketched by children in Mozambique from Kingsley's Lebombo expedition into Vietnam schools. As Vietnam is one of the biggest end-users of rhino horn, it was a key target country for the summit and there is no doubt Trang is a heavyweight. At the age of seventeen, she was named as one of the fifty most influential young people of Vietnam.

Like most passionate people, she can be pretty fiery. When we invited her, she accepted on the spot, but was in London at the time and didn't have a South African visa. She said she would get one right away.

She arrived at the South African Embassy in Trafalgar Square and was told her application would take several weeks to process, which meant she would miss the build-up to the event.

Nobody says no to Trang, and she gave the poor official on the other side of the desk a piece of her mind. Unfortunately, in Africa you need a little more diplomacy than that. Trang's visa application was shoved so far down on the pile that she would probably only arrive sometime next decade.

Trang was crucial to us. She was our link to the key Asian country we desperately wanted to target. We had to get her out here. I phoned Kingsley with the latest drama.

Kingsley has a contacts network more intricate than a spider's web, and it includes the direct numbers of several cabinet ministers. Soon the lines were buzzing. The report back from London was that officials at the embassy were of the opinion that Trang 'needed to learn some manners', or words to that effect. I think Trang would have used stronger words.

This was a major dilemma for us. We had to persuade Trang to keep her mouth closed for a few hours until she got on the plane, not easy with a woman as zealous about conservation as she is. She was still arguing with consular officials when I rang her.

'Trang, just sit tight and don't open your mouth,' I said. 'You are going to get your visa, but keep quiet until that happens. Not a word, understand?'

'OK. I'll try.'

'Once you have it, go straight to the airport and we'll see you in South Africa.'

Fortunately, she listened. She arrived in Johannesburg the following morning and we all breathed a little easier.

With Trang's help, we invited the best of our Rhino Art contestants in Vietnam to the summit. It was crucial for us to show these youngsters the magnificence of a rhino in its natural habitat. They needed to see the connection between the illegal and impersonal sale of rhino horn and that of a vibrant, living creature.

I cannot overstate the importance of this, and one of the overwhelming themes evolving from the summit was that end users' intentions aren't always malicious. It's often more likely to be a lack of understanding of the consequences to wildlife; that a species is in danger of dying out for something of no medicinal value whatsoever. We needed to bring people to the front line to show them the true horror of the situation.

Barry Leitch was co-master of ceremonies with Richard Mabanga, and whenever Barry and Kingsley get together, there is mayhem. I lost count of how many practical jokes were played while we were setting up the site overlooking the iMfolozi valley, but one stands out – particularly as on this rare occasion the person on the receiving end was Kingsley, a world-class practical joker on other people.

Kingsley and his son Ross, who travels with him on all expeditions, said they would organise the campsite for several hundred delegates. That in itself was a logistical nightmare including sleeping quarters, long-drop ablution facilities and a large kitchen to feed hundreds of hungry teenagers.

Kingsley also hired a dome marquee, which would be the summit nerve centre, and four days before the event kicked off, workers arrived to erect the big tent.

However, one of our staff remembered that two years previously the manager of the marquee company, Wally Pelser, had been caught out in a mini-tornado while setting up a venue on the KwaZulu-Natal South Coast. Wally's marquee had taken off like a rocket-propelled hot-air balloon before being blasted inside out. This got Yolande Kruger, our graphics and branding genius, thinking; how could we 'recreate' this to give Kingsley the fright of his life?

Wally gave us a photo of the destruction wrought by the

twister on the mangled marquee, which we put on WhatsApp with the terse caption 'Tornado over iMfolozi'. Everyone on site was then told to turn off their phones.

Kingsley was not at the Hluhluwe–iMfolozi Park that day, and when he got the WhatsApp alert and saw a picture of a shredded tent and hundreds of pounds' worth of damage, he understandably went ballistic.

He then tried to phone us but, as instructed, we had all switched our phones off. That freaked him out even more. He had visions of the entire summit being cancelled and all our work going up in a puff of smoke – or a gust of wind in this case.

When Yolande, tears of mirth streaming down her face, told Kingsley it was a joke, his booming laugh was the loudest of all. However, I have no doubt he is still plotting revenge, dreaming up the mother of all practical jokes to get us back.

The summit opened on 21 September 2014, the day before World Rhino Day, and streaming into the Hluhluwe–iMfolozi Park were 450 delegates from twenty-eight countries, seven TV stations, some of which would be broadcasting live into China and Vietnam, as well as on-the-spot radio feeds. This was absolute gold dust to us. We were speaking directly to the people we needed to most.

The press was magnificent, but reporters are sometimes a mixed bag and you have to think on your feet when dealing with them. In this case, we had a journalist who refused to go on one of the world-famous iMfolozi wilderness trails. There was no way he was going to sleep in the bush, he said with indignation. This was somewhat unusual as he had been sent out to cover a summit concerning the survival of wilderness icons, and this could be a mood-capturing story like none other. Fortunately, his reports on the summit itself were factual.

The three-day programme included cultural sessions, educational gatherings involving people at the front line of the rhino wars and a simulated shootout with 'poachers' involving game rangers and helicopters. This was so realistic that many thought they were watching the real deal. Perhaps it was too realistic, as one youngster wet himself.

A keynote speaker was Ted Reilly, who is a testament to what can be done with iron political will. Ted is CEO and founder of Swaziland's Big Game Parks and a living legend. Thanks to him, Swaziland has the proudest record of rhino protection anywhere in Africa with only three rhinos poached in the last twenty-three years.

Yet it was not always so. In fact, without Ted, there probably would be no rhino left in Swaziland today. From 1988 to 1992, the tiny kingdom lost nearly 80 per cent of its rhino to commercial poachers. The turning point came when Ted found his favourite black rhino butchered and its horn crudely hacked off.

Enraged, he and his rangers loaded the stinking, maggot-infested carcass onto a bakkie and dumped it on the lawn outside the palace of the country's ruler, King Mswati.

Swaziland is an absolute monarchy, one of the few left in the world, and King Mswati has total power. What Ted did was unthinkable and, at the very least, could have resulted in him being exiled from the country he loved, or jailed for a decade or two. The courage it took for him to do that was incredible.

Fortunately, His Majesty was also sick and tired of the near-total elimination of rhinos in his kingdom. He uttered just one sentence, and it was in words his people understood: '*Linyeva likhishwa ngelinye!*' (Put your hand in the fire and you will be burnt.)

After that, Swaziland's Game Act was amended to put suspected poachers behind bars with no bail. Anyone found guilty of poaching or attempting to poach a protected species (rhino, elephant and lion) automatically went to prison for a minimum of five years, which could be increased to fifteen. There was no option of a fine.

On top of the jail sentence, the prescribed value of the animal poached had to be paid to the owner. If the poacher was unable to do so, an additional two years was automatically added on to the jail term.

Compare that to South Africa where poachers are given the option to pay a meagre fine and walk free. Few are put behind bars as the cash they earn from hacking off an animal's horn makes justice a mockery. They happily pay the fines and resume poaching.

The results speak for themselves. In the year of the summit, 2014, South Africa lost 1215 rhinos to poachers. That is more than three a day or one every eight hours, take your pick. In other words, South Africa was losing more rhinos in a single day than Swaziland has lost in twenty-three years.

There were also several youth speakers. One was only thirteen years old, but had already raised thousands for rhino conservation. She got onto the stage, voice tremulous from nerves, then held the audience spellbound.

'Hello, my name is Julia Murray,' she said. 'I'm from Hong Kong.'

She then told her story. She was born in South Africa but her family moved to the Far East when she was three. However, they returned home each year and always visited game reserves. So in 2012, on her eleventh birthday, she decided to skip a traditional candles-and-cake party and instead hold a fundraising event for conservation. She invited other kids over to paint pictures of rhinos and auctioned them off to friends and family, raising over £2000.

Julia decided to donate the money to the Chipembere Rhino Foundation, a conservation project in which Dr William Fowlds – 'who is with us today,' she said pointing to my brother – is intimately involved. She said she had heard about the amazing work Dr Fowlds was doing, how he had saved the mutilated rhino Thandi, and knew the money would be put to good use.

The Chipembere Rhino Foundation is based at Amakhala and was set up by four of our partners, Brent and Chantelle Cook and Paul and Debbie Naude, after the reserve's two male rhinos, Chipembere and Isipho, were poached in 2010. William became a trustee after the Geza tragedy on Kariega the following year. The foundation is dedicated to preserving rhino throughout southern Africa, and the money raised by Julia was used to buy essential radio-tracking collars.

William then invited Julia to come to Amakhala and see what was being done with her donation, and also help collar some rhinos. For the first time, she 'touched the giants' that she loved so much. It affected her profoundly.

'Meeting Dr Fowlds and his team was so inspiring that I knew I had to carry on doing what I could to raise money and awareness to help save these beautiful creatures,' she told the summit audience.

Soon afterwards she created the JuMu Rhino Fund and also won the prestigious Hong Kong Junior Hero of the Month award for her efforts – no mean feat for a South African-born girl in Asia.

She now makes 'We love Rhino' wristbands that she sells for £8. For her, the economics is simple: £4 for materials and £4 for the rhino fund. In other words, it is non-profit in the purest sense. The JuMu fund has so far raised close on £11,000 – which is about 200,000 South African rand.

As she spoke, simply and straight from the heart, I saw Ted Reilly wiping his eyes. Here was a seventy-something, hard-as-nails man of the bush, who had the courage to throw a rotting rhino carcass into the palace of a king, weeping like a child. He was not the only one in tears at this young girl's total commitment.

Finally, Ian Player took to the stage. He could hardly walk and had to be wheeled onto the platform. At eighty-seven, his mind was as sharp as a skinning knife, but his body had been badly broken from the rigours of rhino capture.

In Player's day, rangers had to climb onto the darted animal's head in order to grab hold of an ear to raise a vein and get the antidote needle in. The reversal drug worked quickly – in fact, so quickly that more often than not Player couldn't get off in time so was flicked into the air like a fly by the recovering giant. Player gave not only his heart and soul to these magnificent beasts of Africa, he gave his body as well.

'Wherever you are from, this is your ancestral land,' he told the young delegates. 'You carry Africa inside yourself. This is where humans evolved.'

The crowd was silent as the great man spoke. Richard Mabanga was the *imbongi* (praise singer) for the speech and, dressed in full Zulu traditional regalia, he held the microphone close to Ian's mouth. They were hanging onto every word.

'We are on the way out now,' he said, referring to his era. 'The rhino belongs to future generations. You are the future. It is not an easy job. You have to make your voices be heard. You are the custodians.'

Hundreds of heads nodded in agreement.

Later that day, Dr Ian Player took his final flight over the land that he loved with such intensity. We flew him over the ruggedly beautiful valley where the two rivers, the Black and White iMfolozi, converge.

He died two months later on 30 November 2014. It was fitting in the extreme that his final speech, his final public appearance, was on the soil where the white rhino was first saved. And where the youth of today made a pledge to save it again.

After three days, the delegates went home, inspired by their experience in the wilderness and far better equipped to handle the crisis in conservation that will get even worse on their watch if nothing is done.

These are the outcomes that the youngsters drew up, and were reported by the international media.

To Sound: A world-wide call to action by the youth to save the rhino from extinction

To Send: Key 'hearts and minds' messages from the youth to African and international leaders

To Attract: National and international media and public attention of youth involvement

To Shape: Young leaders to become future global wildlife and conservation ambassadors

To Inspire and Instil: A compassionate concern for wildlife in the next generation responsible for protecting biodiversity

To Increase and Extend: Knowledge of rhino conservation, protection and anti-poaching initiatives

To Showcase: The vital conservation work of members, affiliates and partners of Project Rhino.

It's an ambitious mission statement but, having seen the fire in the eyes of the young over those three days, I have faith they will carry it out. They are the keepers of the flame.

Once the dust had cleared and we were having a few beers around a log fire in the African bush, Kingsley summed it all up in his usual sage way.

'Isn't it amazing that a piece of paper and some crayons led to all of this?'

It was true. Rhino Art started in rural schools near game reserves, a humble grassroots conservation project that sparked the first ever World Youth Rhino Summit.

Perhaps equally unbelievable was the fact that an adventurer-explorer, a dynamo wildlife fundraiser and a former goat farmer in *mblaselo* pants had spearheaded a world gathering attracting potential future world leaders. For three days, the global conservation media had focused primarily on a remote corner of Africa.

It had been incredibly hard work and it was touch and go at times whether it would get off the ground. We had pulled it off, but now we had to get back to our day jobs – back to the rhino wars.

Once again, Kingsley summed it up. 'Guys – we are not in the business of organising summits, are we clear?'

We laughed. Let the full-time organisers handle the next one.

CHAPTER THIRTEEN:

The Horn Market
of the World

What the World Youth Rhino Summit showed was that the rhino wars could not only be fought in the bush. This was not solely a flak jacket and rifle affair.

On the contrary. The battlefield in the boardrooms, corridors of power, university lecture halls and school classrooms was as much a front line for eco-warriors as a firefight with heavily armed poachers in the wild. Probably more so, as that is where you stand most chance of hacking off the serpent's head. In the big scheme of things, the poachers are mere pawns, doing their bosses' bidding for a fraction of the profits.

So for wildlife activists, networks formed at conferences and summits, or contact numbers stored on cellphone directories, were often far more powerful weapons than bullets. There will never be a shortage of small-fry gunslingers in the bush. It was the Mr Big dealers we were after. Things only happen if you knock on doors – and keep knocking until they creak open.

Less than a year after the Hluhluwe–iMfolozi Park summit, our international credentials struck gold when a group of South

African youngsters were invited to visit Vietnam as guests of Operation Game Change, a joint Vietnam/USA government-sponsored initiative aimed at ending wildlife crime in that country.

The culmination of the event was to be a massive outdoor extravaganza called WildFest, which the Vietnamese promised would provide a platform for high-level government representatives, celebrities, business leaders, NGO partners and media to come together and 'show their joint commitment to ending the illegal wildlife trade'.

This was a giant leap forward. More people live in Asia than the rest of the world combined, and the Far East is statistically speaking the sole consumer of rhino horn. And here we were about to engage directly with the Vietnamese government itself. The Vietnamese people were where the message of what was happening to animals thousands of miles away in Africa needed most urgently to be hammered home.

Operation Game Change was organised by Freeland, an international wildlife counter-trafficking organisation, and Traffic, the wildlife trade monitoring network. Although we were invited as a delegation from Project Rhino, we were also asked to do Rhino Art shows as a direct result of the World Youth Rhino Summit's success.

I had just returned from America after a month-long speaking and marketing tour, only to be told not to bother unpacking my bags as I was booked on a flight to Hanoi. My job was to chaperone eight South African youngsters whom we had handpicked out of thousands. Our delegation would not only spread the word, but also perform at WildFest, scheduled to be staged at the Imperial Citadel, the oldest building in Vietnam. I don't think it was possible to have a more prestigious venue in which to deliver a powerful message, particularly one that might not be received positively by all people in that country.

I insisted that Richard Mabanga came with me, and we got on the plane with him dressed in full Zulu *impi* outfit and me in my *mblaselo* pants.

While cruising at 30,000 feet above Asia, we suddenly heard our names mentioned over the Airbus intercom. It was the pilot Captain Conrad Pfeiffer and his South African Airways crew wishing us luck. The entire plane, some 350 passengers, burst into spontaneous applause. We were off to a great start.

Our delegates were mainly sixteen or seventeen years old, and the two more mature ones, Phelisa Matyolo and Nadav Ossendryver emerged as leaders of the group.

Nadav, now in his twenties, is touted as the next Bill Gates and was recently listed in *Forbes* as one of the world's top-thirty young entrepreneurs.

His story is one of dazzling initiative. His parents are avid wildlife enthusiasts and, whenever they toured the Kruger Park, Nadav would pester them to flag down passing cars to ask what animals the occupants had seen. Eventually his mom and dad got tired of this and refused.

Instead of sulking, Nadav had a brainwave and single-handedly created a cellphone app which he called 'Latest Sightings'. Basically, it's a platform for game watchers to report animal sightings and locations in real time. No longer did Nadav's parents have to wave down other cars – all the information he needed was pinging on his phone.

Knowing he was onto something, Nadav started marketing the app. The success was phenomenal. His YouTube channel, known as 'Kruger Sightings', has more than half-a-billion viewers – an African record. The previous record was about 53 million.

Latest Sightings didn't just help Nadav see more animals. From a wildlife perspective, it provided invaluable data on game movements, while sightings of injured and poached animals alerted game rangers and vets.

This was the calibre of youngsters in our delegation. They soon proved their worth, as Operation Game Change was a non-stop whirlwind of activity. Our youth spoke at schools and universities, visited traditional Chinese medicine shops and Asian bear bile farms, judged Rhino Art contests, danced and sang on various stages, and, on one occasion, attended a diplomats' lunch at the US Embassy.

Most importantly, the delegates made several TV appearances, bringing the horror of wildlife crime into the living rooms of ordinary people. Richard and I made sure the youngsters were the spokespeople. We kept in the background.

The seriousness of the message delivered was clear, but for sheer showbiz, there was nothing beating us. I reckon Phelisa's hypnotic dancing, her rhythm, agility and vivacity, had a bigger impact on the people of this vibrant city than anything else from Africa. She and Richard performing together outshone any live show I've seen, even at venues in Las Vegas. And all the while the tempo of the narrative never flagged – 'don't do wildlife crime . . . don't use rhino horn'.

Another highlight was a cycle ride through Hanoi with US Ambassador Ted Osius promoting the slogan 'Don't Accept, Don't Use, Don't Buy'. In other words, have nothing whatsoever to do with any wildlife product. Ted is revered in the country and was one of the first US diplomats sent to Vietnam after the bitter twenty-year conflict between the two countries. In

1996 he helped establish the first US consulate in Ho Chi Minh City – formerly Saigon – and as tensions gradually thawed, he was later personally nominated by President Barack Obama as ambassador.

Take it from me, to ride a bicycle in downtown Hanoi requires nerves of titanium, facing a tsunami of motorbikes that stop for nothing, not even traffic cops. Fortunately, our ride was at 6 a.m. before the *Mad Max* morning rush hour, but I was not sure whether it would be the stifling heat, the pungent two-stroke gas fumes or an errant biker that got me first. Even so, our peloton of cyclists looked impressive in identical outfits with Operation Game Change and Fighting Wildlife Crime logos emblazoned on our gear.

It was also here that I met two extremely brave sisters, Victoria and Vanessa Wiesenmaier, from Cape Town. A year or so beforehand Vanessa had contacted me out of the blue saying she and Victoria planned to do a 5000-mile cycle around South-East Asia to protest against wildlife crime. The tour was called BuyNoRhino, and we agreed that the sisters would also represent Rhino Art. They were in Hanoi to submit an anti-rhino poaching video to the WildFest judges, and it was fortuitous that we were all in the capital city at the same time. The winning film at the festival was going to be screened on Vietnam's mainstream TV, so would have maximum exposure. Vanessa, or Ness as we called her, and Victoria's short movie was great, if somewhat gory, showing a man's nose being cut off in an alleyway. But if that didn't get the message across, nothing would. (Their film was one of the runners-up, a tremendous achievement.)

I spent an afternoon with the two girls and was astonished at their almost insane in-your-face courage. They would fearlessly confront total strangers in the street and challenge them if they were wearing wildlife items, such as ornaments or trinkets.

I watched them go up to one guy who had a tiger's claw necklace. 'Do you know what you are doing wearing that?' Vanessa demanded, as Victoria instantly followed up saying, 'Do you know how much a tiger has suffered for you just to have its claws as an ornament?'

They certainly gave the guy something to think about as he beat a hasty retreat. I was also with them when they interviewed people in the Old Town Hanoi market and they gave animal traffickers hell on street corners for selling caged birds, snakes and lizards.

However, I was worried about their safety and said they had to be more careful. They were tourists in a country with different ideas about equality of the sexes.

They shook their heads. 'We've been here for eight months and know what we're doing,' said Vanessa.

These two eco-Vikings motivated the rest of us with their unbelievable bravery and commitment. They also told me where most of the rhino products in Vietnam were being sold – a village called Nhi Khe about 11 miles outside Hanoi. That got my attention, as I wanted to see for myself the shady side of this great country. Everyone was so hospitable it appeared impossible that beneath the veneer lurked such dark secrets. It seemed inconceivable that many of the amazingly friendly people sanctioned, or at the very least turned a blind eye, to something as satanic as wholesale wildlife slaughter. I knew I had to visit the so-called craft village, and Richard, myself and Nadav's father Ilan made plans to go there and try to buy some black-market rhino horn.

At the end of the two-week campaign, the South African delegation was among those presenting the World Youth Wildlife Declaration to Vietnamese government officials urging that trafficking be stopped. Vietnamese youth leaders and school pupils also signed the declaration, pledging their support to end the purchase and consumption of rhino horn in their country. Once again, the youngsters had spoken.

Our delegates won hearts and minds wherever they went. It wasn't just me saying that. Ted Osius called me aside and said Vietnam will never be the same after the South Africans left. Our youth blew them away; the sheer energy and passion of their speeches and performances were an inspiration to the rest of the activists.

Vietnam is an emerging giant, a potential leading nation of the future. But its wildlife legacy is a stark blemish on humanity. Until it eradicates its insatiable desire for rhino horn, the country will always be tainted.

That was the thought on my mind as I nervously boarded the plane back for South Africa. In my pocket was a Ziploc packet of rhino horn shavings bought at Nhi Khe that I was smuggling back home. I needed it as tangible evidence of exactly what we were up against. I needed to get it tested to see if the animal killed for these pathetic few flakes was from South Africa.

Unbeknown to me, the Wildlife Justice Commission (WJC), an international not-for-profit organisation formed to prosecute wildlife criminals, was busy investigating the Nhi Khe 'craft' market and in the throes of busting it wide open. I had managed to do my undercover work before the commission went public, so my timing was spot on. If they needed evidence, I had it.

Months after I left Vietnam, the WJC issued the following statement:

> During the investigation our investigators observed directly US\$53.1 million in parts and products of rhinos, elephants and tigers trafficked through a criminal network consisting of 51 individuals.
>
> Despite the overwhelming body of evidence, prepared by former law enforcement professionals for Vietnamese law enforcement authorities, an extensive diplomatic outreach and engagement of international stakeholders, the Vietnamese government has failed to take decisive action to close down this criminal network.

The WJC announced it was going to hold a public hearing on wildlife crime at the International Court of Justice in The Hague, Netherlands.

Then they dropped another bombshell. They wanted one of the South African youth delegates who had been to Vietnam's Operation Game Change festival to testify.

That meant we were going to put our case to the highest international court on the planet.

CHAPTER FOURTEEN:

The Hague

Once I had the results back from the Onderstepoort laboratory confirming the horn shavings were southern white rhino, I realised I was sitting on a political hornets' nest.

The reasons were complex. At the time, state conservation bodies such as Ezemvelo KZN Wildlife were in a turmoil as there was concern about possible internal corruption regarding rhino poaching and horn trafficking. The fact that court evidence was sometimes 'lost' and suspected poachers frequently got off scot-free, or with a slap-on-the-wrist fine, did not allay those concerns. As a result, there was an all-pervasive climate of distrust. Anyone could suddenly fall under suspicion for no good reason. For example, a ranger buying a new car could trigger an internal investigation, or else become an innocent victim of a whispering campaign.

So I was told by people I respected that in the current situation it wasn't worth dropping a political grenade providing irrefutable proof that South African rhinos were directly poached for the Far East market, even though it was something we all knew. Now that we had the actual smoking gun – the Ziploc packet in my pocket – it could add kindling to the fire, resulting in more rangers being the targets of malicious rumours. In fact, I could become a target myself, and even be arrested for smuggling.

I reluctantly agreed that the timing was not right. We would instead wait for the best moment to get maximum exposure. I put the Ziploc packet away for another day.

But even though I was not going public, the Wildlife Justice Commission docket certainly was. And with a vengeance. What it contained was dynamite for conservationists.

The 5000-page report resulting from the year-long investigation was the most damning indictment of wildlife abuse ever published. It showed sickening slaughter of endangered species on an industrial scale with alleged connivance of government officials, social media, banks and an almost criminal lack of law enforcement.

Two countries in particular were singled out: China and Vietnam.

WJC investigators claimed to have evidence that in 2015 body parts or products worth US$53.1 million from 907 elephants, 579 rhinos, 225 tigers and other endangered species including pangolin, the most trafficked animal in the world, Asian bear, hawksbill turtles and helmeted hornbills had passed through Nhi Khe.

It was mind-blowing. The 579 rhinos killed represented almost 50 per cent of the total amount poached in South Africa that year, and alone accounted for an unbelievable $42.7 million in just twelve months. The rest was made up of ivory, $6.8 million, and $3.6 million in tiger body parts. No figure was put on the other endangered species.

Also revealed was that apart from selling directly over the counter at Nhi Khe's 'craft' shops – such as my transaction

– social media was the retail channel of choice for wildlife criminals. WeChat, a China-based super-app incorporating messaging, social media and mobile payment, was used almost exclusively to target Chinese clients, while Facebook was used to target buyers in the rest of South-East Asia.

The WSJ said it would also present evidence incriminating seventeen Chinese banks that handled alleged traffickers' accounts.

The Executive Director of the Wildlife Justice Commission, Olivia Swaak-Goldman, said that despite this 'overwhelming body of evidence', the Vietnamese government failed to take effective action to close down the extensive criminal networks.

She said while the authorities had taken 'some steps' to address the over-the-counter trade at Nhi Khe, almost all the key traffickers named in the docket were still active behind closed doors or on social media. Even more alarming was that wildlife crime had moved to other more discreet locations near Nhi Khe. Instead of being stamped out, it was merely relocating.

The WJC bombshell exploded further when the TV network Al Jazeera's Investigative Unit broadcast a video taken by a Dutch group that had gone undercover into Nhi Khe.

In the footage screened around the world, a woman is seen entering a room and openly placing a large rhino horn on a table, while buyers discuss its worth in weight. Also shown are several signs advertising ivory, rhino horn and tiger parts. This smoking-gun evidence was as good as what I had in my Ziploc bag.

The video was now part of the WJC dossier, and investigators claim it would have been impossible for Nhi Khe's shops to operate brazenly on such a large scale without local authorities being aware.

The WJC accusation was clear: the Vietnamese government was at the very least complicit by its apparent lack of meaningful action.

The good news was that China had been more cooperative with the WJC and had started a preliminary investigation, so at least something was being done in one of the deadliest black markets.

I then got a call from The Hague asking us to send one of our rhino youth ambassadors to testify at the tribunal. The WJC wanted to hear first-hand what it meant to a young South African witnessing their national heritage being systemically destroyed by international greed. He or she would be interviewed by the world's top wildlife investigators, and have a global platform to speak about the wholesale slaughter of African rhinos.

The obvious choice was Phelisa Matyolo. She was articulate, highly intelligent, superb on TV and had chutzpah by the barrowload. The problem was that she didn't have a visa and she had to get to The Hague within a week. We were prepared to pull out all the stops and use the same ministerial leverage that got Trang to the World Youth Rhino Summit the year before, but then hit another problem. Phelisa had recently landed a dream job with the South African Tourism Board and was doing an intensive cram-course in Mandarin (she's now fluent). Her bosses refused to let her miss a few days of lectures.

I couldn't believe it. Here was an opportunity to speak on the world's most influential podium and showcase one of the most iniquitous crimes of our time. Phelisa would be directly appealing through a global microphone to the common decency of humanity, and she was told she couldn't go. She was in tears.

I then phoned Nadav Ossendryver, our other top youth ambassador. Nadav has dual citizenship, Israeli-South African, and had a valid European visa. I told him he had forty-eight hours to get to The Hague. He started packing as he clicked off his cellphone and landed in the Dutch capital two days later.

His address was titled 'African Youth Involvement: The Need to Protect Our National Heritage' and he spoke from the heart. He said wildlife was the key international attraction for much of Africa, so the current mass slaughter of iconic animals such as rhinos would have dramatic effects on tourism. This would be catastrophic for many impoverished communities.

Recounting his trip to Vietnam during Operation Game Change, he said the South African delegation was shocked to its core to discover that many people were unaware animals were brutally killed in the 'harvesting' (i.e. hacking off) of their body products. One of his sentences resonated deeply with the tribunal and was prominently displayed on the internet. It tersely sums up the global tragedy we face: 'We found people did not really know what rhinos or elephants were as animals. They were just products.'

Sadly, we also got bad news. Vanessa Wiesenmaier, who had been such a help and inspiration to us with her sister Victoria during Operation Game Change, had died on Table Mountain in Cape Town – far too young, and far too tragically.

I felt incredible sadness when told about it. Words could not do justice to such a brave and inspiring woman.

I wrote this short obituary that was read at her funeral:

Vanessa joined our team for three years in her preparation for their South East Asian Ride. Together with Vicky they took our Rhino Art Programme to numerous schools in Hong Kong, Malaysia and Vietnam.

We then joined them in Hanoi for an incredible two weeks of Operation Game Change and at a film festival in Hanoi, their movie was shortlisted. Ness continued to be an ambassador for all nature-based agendas on her return.

She once said she would struggle to get back to corporate life as it was human neglect that was plundering our planet. She felt she had to make the world a better place for nature.

You will be sorely missed and the BuyNoRhino brand you started, and your incredible character, will be long remembered.

We will do something in your honour.

Those simple words were not said lightly. It's a pledge I am determined to keep.

CHAPTER FIFTEEN:

Rumble in the Jungle

Things were happening at a spectacular pace. In the space of fewer than three years Rhino Art had reached more than a quarter of a million pupils in wildlife-sensitive areas, been part of the first ever world youth summit, and one of our rhino ambassadors had given evidence at the highest international court on the planet.

Then, out of the blue, a project I would never in a lifetime have dreamed about was dangled tantalisingly before me.

It happened completely by chance. Each year I attend the Africa Travel Indaba in Durban, the largest tourism marketing event on the continent, which connects the hospitality industry, tour operators and travel media from around the world.

As wildlife is Africa's most potent selling magnet, no game reserve can afford to miss the Indaba, which means a 'gathering' or 'conference' in Xhosa and Zulu. I go there as a marketing director of Amakhala, promoting our reserve as one of the top Big Five venues on the continent.

While wandering around with a pile of Amakhala brochures looking for people to invite to our stand, I passed an exhibition extolling the virtues of the Democratic Republic of the Congo, or DRC as it's more commonly known. It was eye-catching with colourful posters, while stunning Congolese women in bright West African kaftans and braided hair were sitting behind the stall.

I stopped, wondering why the DRC was at a tourism fair, as it is among Africa's most troubled hotspots. It's the eleventh largest country in the world, spanning a massive 905,355 square miles, but is virtually ungovernable after decades of an oppressive dictatorship. It also has embedded under its vast rainforests the planet's richest deposits of rare earth metals, vital for the high-tech industry, as well as uranium, copper, gold, diamonds and oil. It should be one of the richest countries, yet it is unable to reap the benefits thanks to more than a century of abuse starting with the colonial Belgians when the country was the personal fiefdom of King Leopold II. It finally got independence in 1960, but civil war soon exploded with unimaginable brutalities on all sides. This was during the height of the Cold War and hundreds of mercenaries, mainly from France, Belgium, Britain and South Africa, flocked to the country as guns for hire.

In 1965, the Army Chief of Staff Mobutu Sese Seko seized power through a military *coup d'état* and renamed the country Zaire. He ran it into the ground for forty-seven years, finally being ousted in 1997 by Laurent-Désiré Kabila, who changed the name back to the Democratic Republic of the Congo. Kabila's ascendancy resulted in a second Congolese war involving nine African countries that was even bloodier than the 'mercenary wars'. Estimates are that almost 5.5 million people – mostly civilians – were killed and, to this day, the DRC has the world's largest crocodiles due to protein-boosts from mass feasting on human corpses. Kabila was assassinated by one of his bodyguards in 2001 and was succeeded by his son Joseph.

Given its tormented history, I was wondering what this potential giant of Africa had to offer in the way of tourism. Not much, it seemed, apart from some sport fishing on the Congo River, the world's third largest waterway by volume, where Goliath tiger fish are the targeted species.

I got talking to one of the men at the stand who introduced himself as James Wamboy. As I was the only visitor, I had his undivided attention. With James was Claude Mpassy from the Beatrice Hotel in Kinshasa, the country's only major hotel owned privately by a Congolese citizen.

I was wearing Amakhala-branded clothing and a Project Rhino cap and gave them some brochures about our reserve, not for a second thinking this was going to go anywhere.

I could not have been more wrong. Both James and Claude paged through the brochure with extreme interest. They definitely had something on their minds.

James then put down the brochure, again shook my hand, and said our meeting was extremely fortuitous. He said his boss had acquired a game reserve outside Kinshasa and asked if I would be prepared to meet him. He had an 'opportunity' for me.

'What opportunity?' I asked.

'We want animals,' he replied, looking at my brochure's photos of the wildlife thriving on Amakhala. 'Have you got a contact for us?'

I certainly had. Game relocation and range expansion are among my passions, and I hope never to see any healthy wild animal culled on my watch because there was nowhere for it to go. One thing I did know about the DRC was there was an abundance of untamed wild land, something we needed desperately. The Garamba National Park, for example, is 3232 square miles, larger than some European countries. However, poaching throughout the DRC is rampant, so relocating animals is fraught with problems.

But James had piqued my interest.

'What animals do you want?' I asked.

He rattled off the species they were targeting, mainly giraffe and antelope, and I asked if there were fences around the reserve he had in mind.

'No. We have no fences.'

That could prove a stumbling block, but James insisted that I meet his boss, one of the DRC's most influential entrepreneurs, called André 'Papa' Kadima. In fact, he decided it was so urgent that he picked up his cellphone and then and there dialled Papa. The conversation was in French, but it seemed as though both men were animated. Or maybe French always sounds that way.

Anyway, James clicked off and with a big smile said Papa wanted to meet me as soon as possible. It appeared that what Papa needed as much as animals was a sound business plan, and wildlife project managers are thin on the ground in the DRC. If our meeting was successful, he would employ me as a freelance consultant in restoring and restocking what he said was an iconic wildlife park.

As fate would have it, at the time I was reading *The Last Rhinos*, Lawrence Anthony's final book before he died. Much of the action takes place in the DRC and chronicles Anthony's valiant attempt to save the last pocket of wild northern white rhinos, even meeting with the notorious Lord's Resistance Army (LRA) to plead for a stop to poaching in Garamba. That took courage in spades as at the time the LRA leader, Joseph Kony, was the world's most wanted man after Osama bin Laden.

Although they look almost identical, the northern white is a distinct sub-species from the southern white that we have in South Africa, and when Lawrence was trying to locate them there were only three or four left in the wild. I had visions of restoring a park where we could get those last survivors into a highly protected area and introduce a breeding programme with white rhinos from South Africa to regenerate the gene pool. Maybe we would even use some of our bulls from Amakhala. The prospect of saving a species and continuing Lawrence's amazing vision was to me the ultimate prize. This had the potential to be one of the most exciting conservation projects on the go.

Despite that, I was still a little suspicious that everything was above board. Events were unravelling too fast. My concerns were not dispelled when I received a somewhat tatty *Visa Volante* purporting to grant permission to enter the DRC. It looked a little amateurish to be an official document and was in French, so I couldn't understand a word. I sent it to Lawrence Anthony's widow Françoise Malby-Anthony, who is French, to check if it was genuine.

Françoise called back saying the grammar and spelling was so amateurish that it was definitely a fake.

James and Papa denied this, insisting the visa was authentic and issued by the Congolese government. They assured me I would be safe in their country. I wasn't convinced and had visions of being kidnapped by the Lord's Resistance Army, or Boko Haram in neighbouring Nigeria. Perhaps it was a Congolese version of Nigerian email scams – getting the gullible foreigner into the country to fleece him. Or even being thrown into jail as a smuggler with a fake travel permit.

I just didn't know, and since much of that part of Africa was considered to be the modern equivalent of the Wild West, information was pretty sketchy. For example, I Googled André Kadima and search results came up empty. James had told me that Papa was the DRC's most influential businessman, but if even Google didn't know him, who did?

Still in two minds, I phoned William. He and my parents were not enthusiastic and urged me to wait while we checked out Papa's credentials. I said good luck with that, having just tried my luck on the internet. William also wanted to know where I would be staying in case he had to come and fetch me. His final question was what would the likely ransom amount be. I suspected he was only half joking.

However, Kingsley, who had once been placed under house arrest in the DRC during his many expeditions, was far more optimistic and said I should go. 'Mama Africa will deliver' was his hard-won philosophy, and seeing Kingsley knows more about travel in Africa than anyone else, I decided to heed his advice.

Papa sent me an air ticket, which also looked suspect. So armed with a ticket and visa that I thought might be fakes, I somewhat hesitantly boarded the plane in Johannesburg.

I landed at the N'Djili International in Kinshasa without any fuss. But when I got to immigration, officials pulled me out of the queue and steered me into an office where they took my passport. That was one thing I was warned against – never allow anyone to confiscate your passport when travelling in Africa. It's a golden rule. But it happened so quickly that I was unable to stop them and alarm bells started jangling in my head. Obviously my *Visa Volante* was fake after all and I was about to be flung into a cell.

Thankfully, that was not the case. In fact, the complete opposite. The officials were acting on Papa's orders to look after me and whisked me through the crowded arrivals hall where I was led out into the heat and smog and bluster of Kinshasa.

A first-time visit to the DRC is a full frontal assault on every sensory nerve. It's an Africa I had never imagined, let alone experienced. *Gumba-gumba* jive music blared from thousands of ghetto blasters and I had barely taken a step out of the airport building when at least fifteen people started pawing me, trying to grab my bags, begging to be my porter. They all wanted work and any white traveller is considered a cash machine. That certainly was not the case with me, although I was hoping I would eventually be bringing something far more precious than money to their country. Papa's men fended them off and bundled me into a car.

There is no adequate word to describe Kinshasa traffic. Anarchy is too tame. Traffic officers futilely blew whistles and waved batons to no avail as overloaded jalopies patched with wire and gum bounced along the potholed roads ferrying live-stock and people, some hanging onto the roof. It seemed the only roadworthy vehicles were pristine blue Toyota Land Cruisers belonging to NGOs, UN peacekeepers or aid workers trying to navigate through the chaos.

Even for a sixth-generation African who has ventured into the furthest outbacks of Zululand, this was something way off the charts.

It got worse. As we weaved in and out of the stampede of dented vehicles, one of the guys in the car locked the door. I asked him if it was dangerous outside.

'Not at all,' he said. 'It's to keep the cops out.'

'Policemen? Out?'

'Yes. If the door is not locked, a cop can jump in and you have to pay him to leave. Or else he stages a sit-in.'

Not only that, he said the longer the cop is in the car, the higher the price to evict him.

No cops jumped into our car and sat on my lap, but even so, outside the locked doors was bedlam of the highest degree. Yet beneath the veneer of chaos and madness, this vibrant city is thriving against all odds. There is no unemployment in the informal sector. Whether it's pounding rocks making gravel to fuel the current building boom, filling in potholes and getting a few francs from grateful motorists, haggling over bags of homemade charcoal or decanting tins of two-stroke oil for passing mopeds, everyone is hustling for themselves. No money is ever banked, yet cash is boomeranging around the shanty towns. The hidden economy, unhindered by red tape, bureaucrats or rules, is flourishing.

There is also no serious crime. Moneylenders with bricks of cash operate from folding tables on the sidewalk. When paying for a tank of petrol, a bundle of francs is nonchalantly left on top of the pump.

The most ingenious system of all is buying cellphone airtime. At garages, all you need to do is roll the car window down a fraction to shove through a tatty fifty-Congolese-franc note and read out your phone number. But you only release your grip on the note when the airtime jingle rings on the phone. It's a service that not even a drive-thru at McDonald's has come up with. As always, necessity is the entrepreneurial mother of invention. Due to lack of any other communication systems, cellphones are essential for doing business and there are twice as many in the DRC as there are people. Almost everyone carries at least two.

My 'minders' dropped me off at the Beatrice Hotel, which is owned by Papa Kadima. Papa is a self-made millionaire, who made his money in restaurants and hotels, and the best way to describe him is as one of nature's true gentlemen. He's a grand old man, although not in the best of health and he moves slowly. But even so, you can still sense the energy, see the fire in his eyes that got him to where he is today.

Although he understands English, he doesn't speak it well and often uses his children – all educated in top American universities – as translators. His daughter Olga has a powerful portfolio in the World Bank while sons Freddy, Andy and Dan are in the extended family business.

His wife Beatrice is the impressive matriarch, revered by the family, and Papa's flagship hotel is named after her.

The endearing thing about Papa is he has never forgotten his roots. He's conservative in business, but otherwise exceptionally generous, giving graciously to the poor in hospitals and old people's homes. In short, he's an incredible guy; an entrepreneur, a philanthropist, a philosopher and, as I was soon to discover, a committed conservationist.

My concerns about the trip dissipated. I was starting to enjoy myself. The hospitality at the Beatrice Hotel was superb. This was certainly no confidence trick as I had earlier feared. There was no Boko Haram or Lord's Resistance Army about to kidnap me. It was the exact reverse, as Kingsley, with his advice that Africa is in essence a force for good, had predicted.

Papa has an amazing network. I was soon linking up with influential people and met everyone in the cabinet except President Joseph Kabila. However, Papa once said to me that he never discussed politics at the hotel, otherwise he would lose customers. He is also a wise man.

One event that showed what a big hitter Papa was, was when the DRC won the African Nations Championship, the continent's top soccer tournament. I was in the country in 2016 when they beat Mali 3–0 in the final, and the whole city closed down in wild street celebrations. The team later visited the hotel as Papa knew the coach, Florent Ibengé, a national hero who was being mobbed wherever he went. We spoke to him and he agreed to do work for wildlife in the Congo. To get that type of commitment showed Papa's revered status in the country.

Soon after I arrived, Papa outlined his dream to me. Several years ago he had taken over a reserve forty miles outside Kinshasa that he wanted to revamp into a showcase animal sanctuary. He called the proposed venture the Pride of Africa, and for him it was an all-consuming passion.

The reserve Papa was talking about was Parc de la N'Sele, a 3500-hectare former playground of President Mobutu Sese Seko, the dictator who became almost as famous for hosting the 'Rumble in the Jungle' between Muhammad Ali and George Foreman as he did for bankrupting his people. Despite the fact that Mobutu ruined everything he touched, including his country's wildlife, he loved Parc de la N'Sele and regarded it as his holiday retreat. He built a house, actually more of a palace, which still remains, complete with a massive Chinese pagoda.

The park is named after the N'Sele River that flows into the majestic Congo, surrounded by lush grassland plains studded with beautiful palm trees. The Congo River is more than 6 miles wide at that point, and on a clear day it is possible to see Congo Brazzaville on the other bank. But sadly there are few clear days in the DRC. The country's biggest industry is charcoal, due to erratic electricity supply, and all forests are slowly but surely being scorched. The smoke from hundreds of thousands of bush fires hangs over the country like a sooty shroud, only clearing briefly in the aftermath of a tropical thunderstorm. The sun seldom shines through the murk, but the humidity trapped by the dense smog clouds is relentless.

The Congo River alongside the reserve is the artery running from the country's main harbour, the port of Matadi, to Kisangani, the largest city in the north-east. For most Congolese it's the only route across the country as roads into the interior of this massive nation are usually impassable. The only alternative is to fly, but few locals can afford that.

The river is the lifeblood of the DRC. Without it, the country could not survive commercially, as there is no other viable trade route. In fact, one of the most innovative conservation projects I came across was watching boats funded by anti-poaching NGOs supplying food and goods to the thousands of passenger ferries, thereby preventing the wholesale plunder of animals along the river banks.

During the 1960s and 1970s, Parc de la N'Sele had large herds of wildlife, as well as enclosures for lions, leopards, okapi and chimpanzees, and even an Olympic-size swimming pool open to the public. It has much historical significance, although not all good. It was at the Parc that Mobutu formed his political party, the MPR (Popular Movement of the Revolution) in 1967 and drafted the constitution for his one-party state that gave him absolute licence to pillage the country at will. The document was called the Manifesto of N'Sele.

Ironically, it was also at Parc de la N'Sele that Mobutu lost total power. Facing mounting insurrection in 1991, he held a conference at the Parc to try and establish a government of national unity. Other politicians smelled blood, and that was the beginning of the end for one of history's most brutal despots.

A double irony is that the man who finally ousted him, Laurent-Désiré Kabila, used the road to N'Sele as the axis to march his ragtag army to Kinshasa.

Consequently, the ghost of Mobutu is everywhere and the logo of the hated MPR is still visible on the main building at the Parc. Even though he has been gone for more than twenty years, his brutal legacy bedevils this beautiful country like a demented demon. But still, Parc de la N'Sele is a living piece of Congolese history, and even though it is the country's smallest park, it punches far above its weight in significance.

That was the situation when I arrived. It was a time of cautious optimism. After two civil wars where rival armies fed their soldiers with the wholesale slaughter of wildlife, a genocide in bordering Rwanda resulting in millions of refugees, and poaching of epidemic proportions because 'bushmeat' was regarded as top cuisine, the Congolese Institution for Nature Conservation (ICCN) finally decided to bring sanity to one of the most visited parks in the country. The Parc de la N'Sele. They asked Papa to step in.

Then, completely by fluke at a tourism stall several thousand miles away, Papa's men bumped into me. The question was straightforward – could I handle the project management regenerating Parc de la N'Sele?

It was an incredible turn of events. After arriving in the country with what I feared to be fake documentation, I instead got a top-level mandate: fix the Parc. If I could do that, I could possibly move on to the other much bigger reserves, such as Garamba, Bombo-Lumene and Virunga.

And maybe find the last of the northern white rhinos that Lawrence Anthony had spent the final years of his life trying to save. Although, in my heart of hearts, I knew the animal was almost certainly now extinct in the wild.

CHAPTER SIXTEEN:

Parc de la N'Sele

It was going to be a tough ask. While Parc de la N'Sele had not been totally abandoned since the fall of Mobutu twenty years previously, it had been horribly neglected.

Papa had signed a fifteen-year lease with the ICCN and his team had been managing the reserve for three or four years by the time I arrived. This mainly seemed to involve providing food and shelter for the animals and workers as, by his own admission, he had no concrete business plan. Just keeping the Parc functional with the paucity of skilled rangers and fragmented infrastructure took a lot of effort, and Papa needs to be commended for that.

However, evidence of the Parc staff fighting a losing battle was everywhere. In some areas squatters had moved onto the land, building makeshift huts and planting subsistence crops. Also with the rapid expansion of Kinshasa, a megacity of more than 11 million people, urban creep was lapping at the Parc's borders like a polluted lake. The reserve was barely an hour's drive from the city centre and I noticed with some alarm that already about 1000 hectares of peripheral savannah was permanently lost to squatters and urban sprawl.

This was to be my battleground in the DRC, and one I had to win to prove myself if I was to have any hope of establishing a breeding herd of rhinos in the country. It didn't look good.

Urban sprawl was among the most obvious initial problems. No one wants to visit a game reserve that's being gobbled up by shacks, so it was a no-brainer that we had to erect border fences with extreme urgency.

However, despite everything, Parc de la N'Sele was a beautiful place. Most of it was carpeted by thick savannah with stately clusters of tall palms thriving in the sandy soil, almost like a tropical beach. The winding N'Sele River uncoiled like a giant python through the reserve centre, which made fencing difficult, but I figured not much game would cross the river anyway.

However, I was hesitant in calling it a game reserve as most of the animals were in camps or cages. It was closer to being a hybrid zoo set in the bush. The only animals that appeared to be running truly wild were sitatungas, a swamp-dwelling, amphibious antelope indigenous to sub-Saharan Africa from Cameroon southwards to Botswana. It is an enigmatic creature, uniquely adapted to its environment with splayed hooves to traverse the muddy marshes, and is able to hide underwater for long periods with just the tip of its nose exposed. This enables it to evade poachers, which is how against all odds it is surviving the wildlife slaughter raging throughout West and Central Africa.

But due to their elusiveness, sitatunga are rarely seen and thus not exactly a drawcard on game reserves. We only knew they were in the Parc because we regularly saw spoor. Which probably meant that hunters with garrotte snares would also be after them, providing yet another urgent reason to get the border fences up.

In the cages were gorillas, chimps, baboons, mangabey monkeys, crocodiles and pythons as well as rare birds such as crowned cranes and Congo peacocks. The fenced camps had forest buffalos, zebras, giraffes, ostriches and some emus – which are actually indigenous to Australia.

The animals were not starving, but they were not in peak condition. Some of the big apes that had been cramped in cages since Mobutu's day appeared almost brain dead, possibly through absolute boredom. There was nothing for them to do except sit morosely in their cells.

Yet it was obvious that Parc de la N'Sele had been an impressive reserve in its heyday. There was a media centre, studios, swimming pools and, of course, the ex-dictator's palace, although all were riddled with bullet holes, stark testimony to the country's tormented past. The reserve had originally been built by the Belgian colonists and was phenomenally well constructed. The fact that the buildings were still standing after two lengthy wars and decades of neglect testified to that.

Most of the infrastructure, however, was wrecked or stolen. For example, the prime source of electricity came from a bastardised distribution board that looked as though TV's MacGyver had been there with some rusty paperclips and duct tape. Also, some of the roads were barely passable rutted tracks and the access gates were either open or hanging drunkenly on broken hinges.

Fortunately, there were still loyal staff members tending the remaining animals. Some had been working at the Parc for most of their lives and were incredibly knowledgeable about local wildlife. Sadly, this was not true of the younger employees, most of whom seemed to have little awareness of nature and no interest in acquiring any. They had no idea of animal behaviour or love of the bush, which was unusual to find in people working at a game sanctuary. If any of the animals in the cages 'misbehaved', the younger guards would poke them with sticks and taunt the creatures. I put a stop to that.

There was a baby chimpanzee called Monica who stole my heart. Papa had bought her for $300, and when I first saw her holding on to the bars of her cage while on the floor, amber eyes staring upwards, I told the guards to open the door. She came straight to me with a tight hug, like a worried child clinging to its parent. I spoke to her and played with her, and I think she got more love and attention in those few minutes than ever before in her young life. Her mother had probably been killed in the rainforest for bushmeat, which is how Papa got her, buying her from poachers out of compassion. I never asked him, but that is the melancholic story of most orphaned chimpanzees throughout West Africa.

From that day on, I always brought bananas and Monica would go mad, shrieking with delight. I would swing her around as I had done with my daughters many years previously, walk with her and hold her, and it was traumatic for both of us when I put her back in the cage. She would look at me accusingly. I felt wretched.

Then, as the door slammed shut and I left, she would start swinging aimlessly on the cage bars. I later discovered that if she was 'naughty', as the wardens put it, they would spank her like a child or throw stones at her.

I vowed that one day I would get Monica freed, so she could climb trees, eat wild fruit and be part of a family, living like all chimpanzees should. But Papa considered her to be one of the Parc's main attractions and had no intention of releasing her. Somehow I would have to get around him. I am still trying.

After several days researching the reserve and talking to the more knowledgeable staff members, I started to draw up plans to get Parc de la N'Sele pointing in the right direction as I saw it. Despite my initial misgivings, Papa's determination to make Parc de la N'Sele 'Kadima's Pride of Africa', his undoubted love for the reserve and that he was prepared to open his cheque book to restock game, gave me hope.

My overriding initial objective was to close the cages. For good. That was non-negotiable. I would not put my name on a project where wild animals were locked behind bars.

Unfortunately, this was complicated as some of the animals were too old or infirm to care for themselves. We would even have to consider euthanising them, but there was no turning back. For wild creatures, a cage is worse than a living hell and euthanasia is sometimes preferable. If Papa or the authorities wanted a business plan from me, cages were not up for discussion.

I suspected the issue would be controversial. My suspicions were soon confirmed – vehemently so. Many staff members were horrified and several confronted me to voice their extreme disapproval. Was I considering even releasing the pythons? they asked incredulously.

'Damn right,' I said. 'Even the pythons.'

The next step was a joint meeting with the ICCN who owned the reserve, cabinet ministers who would hopefully agree to my proposals, and businessmen who would – equally hopefully – finance it, and get them to buy into my vision.

One thing I learned in the DRC is you have to look the part if you want to be taken seriously. Smart-casual attire is not an option. Even if you pitch up in big-name designer jeans and a £1000 Italian silk shirt, you would not be considered a big hitter. A tie and jacket is OK, but a tailored suit is first prize.

Unfortunately, suits are an endangered species in my wardrobe, but I managed to shake the mothballs from a navy blue jacket and look reasonably sartorial – by my standards anyway.

Even so, I was the worst-dressed person in the room as every government minister was outfitted in an immaculate set of threads, polished leather shoes and a handkerchief popping out of the top jacket pocket.

I stood in front of the distinguished group and outlined in detail exactly what I planned to do. It was one of the most important presentations of my life.

In essence, the first stage of the proposed revamped Parc de la N'Sele project was the obvious one: constructing a game-proof fence around the reserve and introducing plains animals such as zebra, giraffe and antelope – including rare roan and sable that thrive in savannah-type bush. All rangers would get extensive training with world-class tuition provided by the Field Guides Association of Southern Africa (FGASA).

Next step would be to upgrade the infrastructure, completely overhauling the Parc's roads, the main gate, electricity, plumbing, and improving security.

Then we would upgrade the old Belgian-built multimedia building into a specialised learning centre with emphasis on youth education. This is where my main expertise came in, thanks to my experience of re-wilding Amakhala and the Rhino Art project. It would be called the André Kadima Conservation Centre, with auditoriums, research facilities, wi-fi and generators so we would not be at the mercy of the DRC's erratic electricity supply. Wildlife paintings and landscaped indigenous gardens would complete the pastoral scene, making it a memorable experience for urban Congolese to see the incredible array of animals they didn't even know existed in their country.

In short, the multimedia centre would be a world-class showcase of not only what the country was losing, but what the entire continent of Africa faced if we did not conserve our unique natural heritage.

We would also introduce projects similar to Rhino Art. Kingsley, Sheelagh and I were already far advanced with plans to launch Elephant Art in Kinshasa schools, which was the same concept as Rhino Art except with the forest elephant as the theme. I stressed that youth conservation work would thrive at the André Kadima Education Centre.

My voice hardened when I outlined the next stage. All caged animals able to be freed would be let loose. I repeated that. However, we would tag some of the primates as monitoring their movements would be an indication of how once-caged wildlife adapted to life in the bush. In essence, they would tell the 'story' of species survival in the new Parc de la N'Sele. We would also collar some sitatungas and bring tourists to see these beautiful but elusive creatures.

Once the imported plains animals settled, I would bring in an elephant family using contacts in South Africa to provide not only the animals, but the vets, helicopter and capture trucks. I could sense the delegates' interest pique at this. Few in the DRC have seen a live elephant before, despite the fact that the animals had once extensively roamed both the vast plains and the forests. The remaining herds of forest elephants are all hiding in remote areas of the country.

Finally, I came to the most ambitious concept of all. In fact, it was more than a concept. It was a dream. We would bring in southern white rhinos. To relocate the world's most endangered species into a volatile country was tantamount to throwing bullets into a fire, but I believed it could be done with tight around-the-clock security and heavily armed rangers. If successful, we would start a breeding project – maybe with the last remaining northern whites from Garamba. If they were still alive, that is. This was in 2015 and no one seemed to know for sure, although the consensus was that the species was now extinct in the wild.

There were a few remaining northern whites in captivity, but the species was destined to becoming the largest animal to die out since the woolly mammoth. And here I was talking about a breeding programme, albeit introducing southern white genes. The Parc de la N'Sele project could result in partially restoring a sub-species; that is what could be at stake. Could the prize be higher?

Due to the almost insane risks involved, I stressed we needed President Kabila in person to get behind the project.

There was silence in the room. I paused, worried that I had overstated my case. Maybe I should have been more cautious, considering the sad state of the Parc. But as I stood there in front of a group of immaculately dressed dignitaries deciding the possible eco-future of the largest country in sub-Saharan Africa, five words came to mind.

Go big or go home.

I had done my best.

Then there was applause. Everyone was smiling. I don't think anyone in the room had guessed beforehand how ambitious I believed this project could be. But the main thing was they liked it.

Papa was ecstatic. His dream for the Parc de la N'Sele was at last starting to edge forward.

He called me aside. 'OK, let's start bringing in animals right now,' he said.

'First we have to fence,' I replied.

Papa nodded. I said I would use my contacts in South Africa to get the ball rolling.

I expected him to nod again. He didn't. Instead he grabbed my arm.

'Hold on, Grant,' he said. 'We can get much cheaper fences from China, than from your people.'

This was not what I wanted to hear. Sure, we could get cheaper fencing in China, but not if we wanted to keep wildlife in and squatters out. Fencing a game reserve is a skilled job and has to be done properly. It's not like partitioning off a garden or an amusement park, and can never be done on spit and bubblegum if one is serious about conservation. There were already buffaloes on the reserve and I had outlined plans to bring in an elephant family and possibly even rhinos. I certainly did not want some flimsy chicken-wire barrier protecting endangered species from poachers with AK-47s and machetes.

This was our first clash. It gave me some idea of what I would possibly be up against in future. If we were cost-cutting on this most basic and essential commodity, major issues could be looming ahead.

Fortunately, it blew over. Papa relented and I engaged Francois Wehmeyer, a game farmer from Grahamstown and arguably the best fencer in Africa, to do the work. But it was a Pyrrhic victory. I later discovered Papa's 'it's cheaper in China' would become a refrain in almost all project work in the DRC.

Energised by the ICCN's response to the presentation, I visited the CITES (Convention on International Trade in Endangered Species) office in downtown Kinshasa to get the necessary paperwork completed.

Once again, first impressions were not encouraging. The CITES office was a ramshackle building with a group of people lounging on plastic chairs and a rusty air-conditioner thudding so hard I thought it was going to knock the wall down. To bring in animals required a CITES certificate stating that the importation was legal, which I would present to the ICCN. Once I had that stamped, the next step was to get an import permit authorised by the State Veterinarian. Money changes hands at each step along the way in the DRC, some legitimately, some not.

The import permit would finally be forwarded to the South African State Vet who, if satisfied, would release the animals.

All this could take several weeks, if not months. It's a monumentally tedious job. I'm a conservationist, not a desk jockey, and had no desire to get involved with mind-numbing bureaucracy and mountains of paperwork. That was not what I signed up for. But I suppose someone had to do it.

While all this was happening, I had a stroke of luck. As Parc de la N'Sele is on the banks of the mighty Congo, I couldn't help noticing the staggering number of sawn logs floating down the river to lumber factories. There was no doubt that deforestation was stampeding at an alarming rate and, even worse, much of it was rare hardwood such as teak, mahogany, ironwood and yellowwood. The Congo Basin contains a quarter of the world's tropical forests, second in size to the Amazon. It is one of the planet's most crucial green lungs, vital for global survival, so I got talking to Masupa Kambale, the Kinshasa representative of the International Tropical Timber Organization (ITTO) to voice my concern. The ITTO, which is based in Japan, is in charge of international tropical forest sustainability, decreeing which trees can be harvested and which can't.

When I told Mr Kambale what we planned to do with a sophisticated multimedia centre at Parc de la N'Sele right next to where logs were being ferried, he was extremely interested. Coincidentally, the ITTO was also setting up education centres promoting forest conservation, and Mr Kambale said he would speak to his head office in Tokyo.

I soon had my answer. The ITTO agreed to donate $450,000 to our media centre. The education side of the proposal – a key component – was now financially secure.

That was a massive weight off my shoulders. Now I could concentrate on what I loved doing, sourcing animals for the park, starting off with zebra and antelope. And, if all went to plan, elephant and the ultimate prize – rhino.

Everything was slotting into place perfectly. Perhaps too perfectly. Despite my initial successes, I was slowly learning that everything in the DRC is done by lobbying rather than consensus. A simple donation can turn into a complex interpretation, and not always what was initially intended. Leverage was an essential bargaining chip. Progress was sometimes sideways rather than forward, and it was not always tangible, let alone definable.

That was one issue that kept nagging. I was charging ahead, thinking I was making headway speaking to the government, CITES, the ITTO and local dignitaries, but a small nagging voice in the back of my head kept chirping away – did all role players have the same vision? To coin the old cliché, were we all on the same page?

More importantly, would I make a difference for conservation in this amazingly vibrant but totally chaotic country?

I didn't know.

CHAPTER SEVENTEEN:

The Heart of Africa

Part of the small print in the Parc de la N'Sele deal was that I would also look at helping manage the Kinshasa zoo, something the cash-strapped ICCN was also desperate for us to take over.

However, I was so wrapped up in drafting plans for the reserve and running corporate chores for Papa during the initial visits that we never got around to it.

The zoo's full name is the Jardin Zoologique de Kinshasa, and consists of a complex of dilapidated cages and compounds in the middle of a man-made forest slap-bang in the centre of the capital. As the ICCN had no money to run it, they had visions of Papa being the financial fairy godfather. But Papa first wanted my opinion before committing himself, so on my third or fourth visit a retired park official, Papa Gerard Ipantua, took me down to the city centre to have a look.

It was depressing. In fact, it was heartbreaking. To call the zoo chaotic is being kind, despite the efforts of the staff, many of whom were doing their best in a dire situation.

In its heyday, the once-showpiece menagerie was considered a blueprint for zoos around the world. Today, it is little more

THE HEART OF AFRICA

than an animal shanty town, a tumbled-down assortment of unkempt enclosures housing emaciated creatures that look as forlorn as their surroundings.

Built in 1933 by the Belgians, for many years it also served as a transit zone for animals destined for European zoos where any wildlife from Africa was guaranteed to 'wow' crowds who had seldom seen such magnificent or exotic creatures. After independence in 1960, the ICCN took control of all the country's conservation projects, including the zoo.

At its peak, the zoo housed 3000 animals. Today it barely has 150, mainly primates such as monkeys, chimpanzees and a few gorillas. There is a leopard and a few civet cats, as well as various species of antelope, wild birds and reptiles such as pythons and crocodiles. Most come from the Congo River Basin and, for many of the city's residents, it is their only tenuous link with the wild creatures of their country.

The director, Dr Arthur Kalonji, was desperate for us to get involved as he had to run this massive complex on little more than wire and gas fumes. He's totally dedicated to the almost impossible job before him and told us if he had money he would bring in lions, which would get the number of visitors up and increase gate takings substantially.

I knew what he meant. I saw groups of schoolchildren visiting the ramshackle complex, but there was not much for them to do once inside. Such outings are usually filled with excitement and boisterous kids, but at the Kinshasa zoo there was nothing for anyone to get thrilled about with miserable, malnourished animals slouching in cages with roofs about to cave in.

Coming from the wide open spaces of Africa, I'm not a big fan of zoos but a lot of them do incredible work in the big picture of conservation. For example, the San Diego and Los Angeles zoos have enlightened millions of kids and spawned generations of animal lovers through the educational work they do. There is a fine line between virtue-signallers who denounce zoos, and those who grasp the reality of what is actually happening on the ground in the battle to save the planet. Zoos are also a repository for gene pools and at the time I was inspecting the one in Kinshasa, the only three northern white rhinos we knew to be alive were in zoos in the Czech Republic, about to be relocated to Ol Pejeta Conservancy in Kenya. Without zoos, we would probably not have any northern whites left.

But Kinshasa's zoo was light years removed from the educational centres you get in the more affluent world. Most of the food came from the generosity of visitors who bought mangoes and bread from street vendors to throw into the cages. A charity also collected leftovers from Kinshasa's restaurants, which is better than nothing but not exactly compatible with a wild animal's diet.

I later met Papa at the Beatrice Hotel and shook my head. The zoo was too far gone, in my opinion. There was nothing we could do to save it. Throwing money around would be little more than a short-term sticking plaster. What was needed was a sustainable business plan, complete infrastructural upgrade, proper nutrition for the animals and more humane cages and enclosures. That could only happen with long-term international aid from multiple donors and investment.

In other words, it was beyond the scope of a private individual, even someone with deep pockets like Papa. Worse, it would distract us from our main goal, which was to make Parc de la N'Sele Papa's Pride of Africa.

Papa understood my reluctance and agreed that we concentrate on N'Sele.

I was now based in the DRC for at least a week every month and the Beatrice Hotel became my second home. Papa also was travelling extensively, particularly to America where some of his children are based, so he was often away when I arrived. However, he had an endless list of stuff he wanted done. I would frequently start the morning dressed in jacket and tie, shaking hands with cabinet ministers, then changing into work-worn khakis and sweating it out in the tropical heat measuring fences.

With Papa often away, I was frequently in contact with the ICCN on his behalf and eventually I was asked to survey some of the country's other parks. As with the zoo visit, I was always chaperoned by Papa Gerard Ipantua. He was invaluable as, although he spoke broken English, it was infinitely better than my French. As a highly respected retired park official, he also knew everyone in Congolese conservation.

I suspected the ICCN thought I had loads of international money backing me, which was useful even if not true. I never confirmed or denied it, and as a result had a unique opportunity to see the DRC's spectacular wildernesses with the blessing of the highest government echelons. In this I was helped hugely by the international group Wildlife Conservation Society (WCS), which is doing a wonderful job in the country, ably led by Richard Tshombe, who provided me with free flights anywhere in the Congo Basin.

There is no doubt that the DRC is a land of almost un-imaginable possibilities, and one reserve that grabbed my attention was the Réserve et Domaine de Chasse de Bombo-Lumene, not far from Kinshasa.

From the moment I arrived, I was captivated. This is some of the most beautiful wild land in the world, bordered by the confluence of the Bombo and Lumene rivers. This meant that much of the reserve did not have to be fenced as the rivers formed a natural barrier. The water was so clear that I could see the glistening bottom 5 metres below as translucent as a Caribbean lagoon and teeming with fish.

The reserve itself was a biodiversity dream, 300,000 hectares of rich savannah plains fringed with woodlands of rolling hills. Rocky kopjes provided nests for birds of prey, and from the top of craggy cliffs the horizons stretched as far as the eye could see. Anthills towering like church spires dotted the plains and, with every breath, I inhaled the dust and beauty of ancient Africa.

For a tourism venue, it was unrivalled. It had everything going for it. Apart from being about a hundred times larger than Parc de la N'Sele, by DRC standards the basic infrastructure was good. It was only a two-hour drive from Kinshasa's airport on a reasonable road that circumvented the chaotic city traffic.

Yet . . . when I say it had everything going for it, I am not being completely truthful. There was one thing missing. Animals. And that, sadly, is the story of the DRC.

At one stage there had been thousands of buffalo roaming the plains. Today only fifteen are left. Once there had been so many hippos that at places the rivers were clogged like a traffic-jammed motorway. No longer. They have been shot out for their teeth, which is now the main substitute for elephant ivory. The remaining animals are so wild and deep in the bush that they are impossible to find. All you see is an occasional spoor or dropping.

I met the park CEO Fabius Monya at the reserve's bullet-pitted entrance sign where the admittance formality involved scrawling your name in a school exercise book. His cheerfulness belied the insurmountable tasks he faced each day, patrolling 1160 square miles with just twenty-four game guards. Most of it was done on foot as he only had one truck and there was seldom money for fuel. His game guards came from the Batwa tribe – sometimes referred to as 'pygmies', a term some tribes regard as insulting – and they were armed with obsolete bolt-action rifles. I doubted many had bullets.

The ranger's biggest concern is not poachers, as so many of the animals had been shot out, but charcoal producers. Villagers are chopping down the reserve's priceless forests to burn for charcoal. It is happening on an industrial scale as charcoal in many areas is the sole fuel source. To get a grasp of the extent of the destruction, all one has to do is log onto Google Earth and call up a live satellite image that shows fires raging every-where. Some burn for weeks on end, decimating grove after grove of irreplaceable hardwood. For many villagers, selling charcoal is their only income, so it is a complex problem.

On my second visit I bumped into a tourist, a Dutchman travelling through Africa in a camper van. I think we were equally astonished to see each other. He said at US$30 a night it was still a magnificent place to camp, despite the lack of animals and facilities.

My mind was sprinting off in tangents, conjuring up visions of what Papa and I could do with this chunk of crown jewel Africa. We would fence where necessary, restock with both plains and forest animals, grade a light-aircraft runway and build a five-star lodge on the pristine river banks, while keeping the 'back to nature' ethos that my new Dutch friend enjoyed.

In short, this had the potential to be a wilderness reserve in a million.

After a couple of visits, I drew up a detailed development proposal unlocking Bombo-Lumene's unlimited possibilities and showed it to both the head of the ICCN, Revd Dr Cosma Wilungula and Papa Kadima. Revd Wilungula loved the plan and I really thought I was getting somewhere – but Papa was hesitant. I was used to his 'it's cheaper in China' comments whenever we discussed costs, but for the first time he spoke of a tighter overall budget. He told me he was financing everything with surplus cash from the hotel and did not have unlimited funds. He had already ploughed a lot of money into Parc de la N'Sele.

This brought me down to earth with a jolt. Perhaps Bombo-Lumene had spoiled me, but I couldn't help wondering whether we had our priorities right. Considering the vast potential of the DRC, was I correct in concentrating on the country's smallest reserve at N'Sele?

Our strongest asset at Parc de la N'Sele was the education project, reaching out to a population with little environmental awareness. But the difference in scale of what we could achieve on 3500 hectares with encroaching squatters and urban sprawl compared to 300,000 hectares of pristine wilderness – once the charcoal producers were cleared – was immense.

Papa did to some extent agree, but told me to get Parc de la N'Sele right before starting on anything else. He was the boss, and that was what I was being paid to do. I got on with the job.

Then in August 2015 Kingsley Holgate arrived, leading another of his Mama Africa expeditions. It was tantamount to a hurricane hitting the DRC – he fizzes with sheer energy and exuberance.

This time Kingsley's lofty quest was to find the geographical centre of the continent and the expedition was appropriately named 'The Heart of Africa'. Using the 'centre of gravity' method to determine Africa's midpoint – the same method used to determine the topographical hearts of Australia and the USA – the International Geographical Union worked out that the most central point of Africa is 17.05291°E, 2.07035°N. The exact spot is in impenetrable rainforest south-east of the uninhabited Nouabalé-Ndoki National Park in Congo-Brazzaville and virtually inaccessible. In fact, Africa's 'heart' has never been surveyed on the ground. There is no signpost, no rock pillar or crude stone cairn to mark it. Even *National Geographic* magazine describes it as 'the last place on earth'. There are few – if any – wilder or more desolate places on the planet.

To say this was right up Kingsley's street is like observing night follows day. And he planned not only to find it, but to empty a traditional Zulu calabash filled with spring water from the Cradle of Humankind at Maropeng in Gauteng, South Africa, where hominid fossils dating back 3.5 million years have been found. It would be done on the exact satnav coordinates as a symbol of hope for the continent.

I was in Kinshasa at the time, so we made plans for an epic promotion of both the expedition and the Parc de la N'Sele project. Kingsley told me that the 'heart' lay in one of the most important strongholds for wildlife, surviving only because the thick forest was so impenetrable for humans. Thousands of elusive lowland gorillas lived there alongside endangered pygmy elephants, forest leopards, African golden cats and chimpanzees – 'many of which have never seen humans,' he added for emphasis.

This conjured up an interesting image. I could picture a baffled chimpanzee's reaction to its first sighting of a human in the heart of Africa – a sun-bronzed giant with steel-grey beard and shoulder-length hair emerging from the tangled vines and undergrowth shouting '*Jambo!*' That's the Swahili word for 'hello'.

Kingsley's son Ross, as much an adventurer as his father and the expedition's navigator, was a little more circumspect about their chances.

'To reach the geographic centre point of Africa we will need Land Rovers, mountain bikes, river boats, dugout canoes, our own two feet, and probably elbows and knees.'

As with all Kingsley's expeditions, it would include humanitarian aid such as distributing mosquito nets, water purification filters, spectacles and community conservation education.

The conservation angle was Elephant Art, with Papa being the DRC patron. There was no better-connected person in West and Central Africa for the job.

The expedition would only stop in the DRC for several days as Kingsley's team was crossing the country at its narrowest point between Angola and Congo-Brazzaville. The fact that it coincided with the announcement of the Parc de la N'Sele project was perfect timing.

On paper, it was a great idea – two visionary ventures coming together. Papa started arranging celebrations at the Beatrice Hotel, inviting most of Kinshasa's VIPs along with TV crews and the press. The tables would be creaking with the finest cuisine and French champagne, and we geared up for the party of the year.

There is a saying 'Men plan, God laughs', and nowhere is this truer than in Africa. We soon discovered that while

Kingsley's expedition was hailed as a stunning success wherever it went, there was no guarantee that would be replicated in the DRC.

The fiasco started at the border where Kingsley's crew, consisting of himself, Ross, Mike Nixon and Bruce Lesley, was delayed for several hours by paper-shuffling bureaucrats. Then, when it was their turn to be served, customs officials closed the door saying it was 'lunch hour', which put Kingsley's team even further behind schedule.

Eventually, after filling in endless forms on which Kingsley stated he was a priest, Ross a veterinary surgeon and Bruce Lesley an astronaut, they were allowed into the DRC, only to be gridlocked on roads that, in Kingsley's words, 'hadn't seen a grader since the days of Mobutu'.

Then they were on the 165-mile 'hell run' between the port of Matadi and Kinshasa. They had to do this section as darkness fell, which can be life-threatening. Even Kingsley, the most experienced Africa traveller in the world, was alarmed as overloaded charcoal lorries without headlights and suspension springs flattened to the axles skirted from side to side, making overtaking suicidal.

Even worse, unlit broken-down trucks scattered along the side of the road created lethal hazards only seen at the last minute. The delay that had started at the border was now a freefall chain reaction.

The party at the hotel was scheduled to start at 7 p.m. We could not contact Kingsley, so had no idea that our esteemed guests of honour were risking life and limb in a mad dash through traffic anarchy.

For five long hours, Papa and I tried to keep the dignitaries entertained. But as Kingsley was the star of the show, we were unable to start the banquet and open the booze until he arrived.

Eventually we decided to hell with it and Papa and I handled it ourselves, conducting press interviews, thinking on our feet as we didn't have any prepared speeches, and just winging questions as they came. It was a flop of epic proportions.

The expedition team finally arrived at midnight. Despite this, we managed to get some TV coverage but even that was pure Monty Python. Or perhaps, in this case, Inspector Clouseau. Very little English is spoken in the DRC, so when Kingsley was interviewed, another expedition member, Deon Schurmann, a 130-kg hulk who played professional rugby in France, was pressganged as a stand-in translator. However, instead of inserting Deon's translation as a voiceover, the programmers simply embedded the audio and muted Kingsley's English. As a result, the world's most intrepid explorer ending up speaking French in a gruff Afrikaans accent with lips moving totally out of sync to the soundtrack.

Kingsley's team were scheduled to leave the day after the DRC's initial Elephant Art launch at St François College, a Catholic convent in Kinshasa. But the debacle continued to plague us as the car ferry to Brazzaville broke down. There was no other way across the vast Congo River.

Even Papa, one of the most influential men in the DRC, couldn't pull strings, particularly as there is some hostility between the two countries. Consequently, the two scheduled nights at the Beatrice Hotel extended to three, then four ... and eventually the staff were referring to the expedition team as residents rather than guests.

Papa loves Kingsley and there was no one better than he to show the team around Kinshasa as they waited for the ferry problem to resolve itself. He took them to the slums, the old people's homes and the teeming maternity wards where he sponsors unheralded humanitarian work for no other reason than these are his people. His charity is truly from the heart.

He also showed them the flip side, that life in the DRC is not all grind, grit and poverty. Kingsley, as always, eloquently summed it up. 'At the plush Beatrice Hotel we experience the enormous gap between rich and poor when Papa invited us to a wedding party. Wherever one looked there was Moët champagne, Hugo Boss suits, pointed shoes, eau de cologne, gold cufflinks, live music and a fashion parade of beautiful girls; not to mention the Lamborghinis, supercharged Range Rovers and stretched Hummers in the car park.'

This was a far cry from most people's perceptions of the DRC and Papa effortlessly strides both sides of the social spectrum. To not understand that is to not understand Africa.

Finally, at exorbitant cost, completely off-budget as most of Kingsley's expedition money goes to charity, we managed to track down a private barge owner with a tug that probably first saw service under Blackbeard the Buccaneer. But even more exorbitant was the bill presented by the Beatrice Hotel. The expedition had stayed there for the best part of a week at $400 per person a night, not to mention that they had some memorable parties.

Kingsley took a deep breath. There was absolutely no way his shoestring budget could have anticipated a bill for an extended stay at a five-star hotel. It was a crippler. The expedition was over.

For most people, that is. Not Kingsley. He then did what every street-smart African would do. Negotiate. He told Papa that he could not pay. Instead, Papa must come to South

Africa and Kingsley and I would refund that hotel bill with undreamed-of hospitality. It would be repaid with interest.

Papa, bless him, paid.

But, unbeknown to us, the storm clouds were gathering.

CHAPTER EIGHTEEN:

Tumble in the Jungle

Despite the 'Heart of Africa' expedition's debacles in Kinshasa, Kingsley and his team managed to be the first explorers to reach the geographic centre of the continent.

It was incredibly tough going, taking seven days to trek the final 10 miles. At times they would half walk, half crawl for five scorching hours to cover a paltry few hundred yards. Kingsley's expeditions are so rooted in adventure and fun that most people are not aware of how dangerous and gruelling they actually are. He was sixty-nine years old at the time, and there he was hacking his way through sweltering rainforests and sucking swamplands in the harshest conditions possible. To call that hardcore is understating the case.

When they first entered the human-uninhabited rainforest, he asked the expedition's Ba'aka porters what to watch out for. The Ba'aka, another tribe who are referred to sometimes as 'pygmy', put up three fingers and their spokesman translated in French: angry forest leopards, angrier forest elephants or even angrier *Ndolo* – the spectacular fawn-and-yellow Gaboon viper that camouflages itself as a litter of fallen leaves. It's one of the most toxic snakes in the world, and a bite that far out in the jungle would be fatal. No question about it.

Kingsley later admitted that after a week of 'grabbing roots to pull ourselves on our bellies through muddy goo, constant wading, sweat, bees crawling up our noses and blood pouring from torn skin', the team almost called it quits.

The final mile was the worst, taking them seven hours – about 180 yards every sixty minutes. The porters, through their interpreter Nazaire Massamba, told Kingsley that it was only the 'wild determination' of the white men that kept them going.

Not only was the expedition another first for Kingsley, the team had also brought the concept of Elephant Art to Central Africa, and I was considering piggybacking on this to include Gorilla Art.

My main focus outside the Parc de la N'Sele project was Rhino Art, and a month or so later I was in northern KwaZulu-Natal on a tour with Richard Mabanga when my cellphone bleeped. It was at night and I was far out in the bush, so it took me by surprise.

It was Papa. He sounded pretty agitated and told me I had to get to the DRC right away.

I told him I was miles from civilisation, but he was adamant. The South African Minister of Tourism, Derek Hanekom, was arriving in Kinshasa the next day and Papa said he needed me to brief the delegation on what we were doing at N'Sele. The fact that I was a South African could add extra impetus, or so Papa assumed.

He said he was transferring the airfare costs into my bank account as we spoke, and would email the *Visa Volante* which I could pick up on my cellphone. He repeatedly stressed it was absolutely imperative that I was on the next flight to Kinshasa.

I jumped into my car, phoned Angela and asked her to meet me on the motorway to Durban's airport with a suitcase of clean clothes. Just before midnight I was in Johannesburg boarding a South African Airways flight to the DRC.

I arrived in Kinshasa early next morning, and several hours later was seated in a plush office with Hanekom and his counterpart, the DRC Minister of Tourism Elvis Mutiri wa Bashara.

I think Papa wanted to show Hanekom that Parc de la N'Sele was going to be the new face of the country's conservation. With my background of re-wilding Amakhala and bringing Rhino Art to hundreds of thousands of schoolchildren, he decided I could vouch for the Parc's credibility.

Tourism in much of Africa revolves around wildlife; it's the continent's crown jewel, its flagship – its soul. Without the wildernesses, the spirit of the land would shrivel like a charred corpse. So meeting with the tourism ministers of two of the continent's giants could be hugely significant for Papa's dream reserve. We were right up there on the radar.

I'm not sure how much we contributed to the top-flight ministerial discussions, but the end result was a five-year agreement to strengthen tourism links. It was great news. We were now on the final lap with both my government and Papa's mandating closer travel ties.

Things were looking good. Maybe my vision of breeding rhinos at Parc de la N'Sele was not so far-fetched after all.

Being in conservation is sometimes like bungee jumping. One minute you are standing on top of a bridge, the next you are plummeting down a canyon hoping like hell the cord doesn't snap.

The first I knew that I was plunging off the canyon wall was when Papa somewhat obliquely told me I no longer needed to travel to the DRC so often.

'Why not?' I asked. I thought we were about to enter the busiest stage of the N'Sele project – the upgrades and new construction work.

'There are problems.'

I asked what sort of problems.

'Congolese problems,' he said. 'Nothing you need to worry about.'

He told me instead to concentrate on getting the fencing sorted out and sourcing animals – which I had already done. Francois Wehmeyer and his team of fencers were on standby, waiting for money to be transferred to various accounts before flying to the DRC, and the first consignment of Burchell's zebra was about to be herded into quarantine pens.

Then Papa, somewhat puzzlingly, told me to stop everything and wait for a while. There was obviously something going on and he did not want me in the loop.

This was way out of character for him. He usually phoned at least ten times a day with rapid-fire questions about buying animals, CITES documents or other wildlife queries. Now there was silence. When I spoke to his assistant, James Wamboy, he said Papa was 'distressed' as some of his family did not share his love for conservation and were either advising him to ditch the N'Sele project or hampering progress. They did not understand that it was his dream, not a hobby. These 'problems' were not good for Papa's health, said James.

The next time I was in Kinshasa I cornered Papa at the Beatrice Hotel and asked what was truly going on. He initially replied that I was like a son to him and should not worry. But after a few beers he opened up.

The current difficulties, he said, stemmed from the Congolese government and were mainly bureaucratic stonewalling.

'Never do business with governments,' he said. 'You never know where you stand.'

This surprised me. I thought the ICCN was on board with the Parc de la N'Sele project. The Director General Revd Dr Wilungula had even asked me to look at other reserves such as Bombo-Lumene. Indeed, Papa and Revd Wilungula were friends.

I soon realised this was a classic situation in African politics. It is not personal. It's survival. Political infighting often does not affect friendships, unless it's a leadership struggle resulting in bloodshed. It's just politics.

From what I gathered, some officials suddenly got cold feet. They were not that enthralled with the good publicity the Parc de la N'Sele was getting. A viable business plan was already in place with a major cash injection from the ITTO, and the South African Minister of Tourism had signed a deal with the DRC. Most people would consider that good news. But not the ICCN, it seemed. If a private project was too successful, some officials were concerned it might embarrass the government.

Papa said it was a problem I could not solve, which was why he hadn't told me before.

Unfortunately, he didn't tell me that because of all this he had also decided to pull the plug on the entire N'Sele project.

It could not have happened at a worse time. Firstly, the fencing for the reserve had arrived in the DRC, but Papa then cancelled the deal with Francois to erect it. So I had some understandably angry people on my hands.

Next, equally alarmingly, the animals I had ordered were languishing in quarantine pens with nowhere to go. I had placed orders on Papa's behalf for zebra, giraffe, antelope, kudu (a species of antelope) and wildebeest, but as he was not making much effort to have them flown out, it was rapidly evolving into a logistical nightmare. The deposit we had put down was being gobbled up in holding-pen costs and animal feed. Almost overnight, I was facing several simultaneous crises.

I had no idea what was going on. I dreaded getting phone calls in case it was the State Vet or the fencers asking what the hell was happening. I had no answers. Only later did I discover the real reason for the delays and soaring quarantine costs was because Papa was searching for more land to create his Pride of Africa dream.

Even worse, with bills skyrocketing, Papa suddenly got it into his head that I was ripping him off. It was a stupid misunderstanding, because when I quoted animal prices, for some reason he thought the figures included transportation. That was not the case. Unless specified, in the wildlife trade all transportation costs are paid by the new owner. Simple as that. And if the animals are being taken to another country, the new owner has to pay quarantine fees as well.

In fact, transportation is usually the single biggest cost. It may cost about $50,000 to buy eight zebras, but hiring a Boeing 747 cargo plane to fly them out costs in the region of $150,000. Somehow that got lost in the translation and Papa bizarrely thought I was trying to make extra money on the side, even when I showed him receipts.

In the middle of all this uncertainty, Papa came to visit South Africa as part of the Beatrice Hotel deal that Kingsley and I made after the ferry fiasco in Kinshasa.

We vowed to give him the holiday of a lifetime, and we did. We took him to the best game reserves in Zululand, he stayed at Kingsley's magnificent house on the KwaZulu-Natal coast, and then at Amakhala, where my father Bill regaled him with stories that had both men rolling off their bar stools with laughter. Papa even took to calling my father 'Dad', which was a bit unusual seeing as Papa wasn't that much younger. It was an eye-opener for him to see how professionally the top Big Five reserves operate in South Africa, and after that we took him to Cape Town. Like so many tourists, he fell head over heels in love with arguably the most beautiful city in the world. Except, being Papa, he took it one step further and bought a smart villa in the affluent suburbs, and he still visits regularly.

It wasn't all play and no work, and I took him to see his animals dying of boredom in quarantine. He still carped at what he considered to be exorbitant fees, but at least he now understood that the actual price per animal was often the cheapest part of the deal. There was no longer any weird suspicion that I was syphoning off funds.

Before he left, he dropped another bombshell. He had finally bought a chunk of land outside Kinshasa near the village of Kimpoko, where he planned to build his new park. It would be called the Pride of Africa Kimpoko. He wanted me to fly out and have a look at the new venue.

At least we now had an address to which to send the confined animals. However, of all the animals I ordered for Papa, only eight zebras reached their destination. They had been in pens for almost nine months, the longest quarantine period in history according to the State Vet. We bought them as foals because smaller animals are cheaper to export, but they were almost adults when they finally arrived in the DRC. I know how they felt – I had aged in animal years myself during the entire exasperating process. Interestingly, when I was at the airport finalising the export formalities, there was a consignment of seventy-one young giraffes being flown to reserves in the Middle East. The demand for wildlife is fast becoming one of Africa's biggest export markets, and it strengthened my determination to concentrate on relocating animals within our own continent. Although habitat encroachment is exploding, the DRC has shown that we still have vast empty spaces to restock with Africa's incomparable wildlife.

One thing Papa hadn't bargained on was the $450,000 grant I had secured from the ITTO being withdrawn. He thought the money would conveniently be siphoned off to the new Kimpoko reserve, but I warned him that might not be the case.

Sure enough, the ITTO refused to re-route the money and the deal was cancelled. I saw their point. They wanted to invest in a project with a detailed education programme, part of which would be outlining the vital need to preserve the DRC rainforests. If that deal was not going through, they had no intention of signing a blank cheque. It had been pre-approved in Japan and any changes, they said, would be a breach of contract.

This was a big blow to Papa, but he went ahead with his Pride of Africa vision and I flew out to have an on-site inspection.

The reserve was on a hill about 20 miles from Parc de la N'Sele, and very small – only about 350 acres. It had some natural bush and waterways but it lacked the spectacular beauty and vista of the imposing Congo River that was part of the attraction of Parc de la N'Sele.

By the time I arrived, Papa had moved animals that belonged to him from N'Sele, including the baby chimpanzee Monica that had so captivated me. It broke my heart to see her still in a cage.

At least she was in a bigger one, and there was no doubt she was better off than she had been at N'Sele. But a cage is a cage.

As always, Monica leapt around in excitement when she saw me, and fortunately I had remembered to bring bananas. But it tore me apart. I felt awful when it was time to put her back in her cell. Monica was one of the reserve's most prized attractions and could be handled by children, so there was still no way Papa would let her go. He would never sell her to me. Barring a miracle, I sadly had to accept that it was unlikely I would be able to set her free as I previously pledged.

Kimpoko also had some free-ranging wildlife such as buffalo, zebra, sitatunga, duiker (another type of antelope) and other natural species, but this was still a far cry from wild Africa. It reminded me of a theme park, even more so when I saw donkeys and peacocks strutting around the buildings.

Papa took me aside and asked what I thought of his new reserve. I shrugged non-committally and said it had some potential. One of the advantages was that many of the schoolchildren who had visited Parc de la N'Sele were now coming to Kimpoko, so educational projects such as Rhino Art could be very successful. Even though we lost the ITTO money, which was initially earmarked to establish an ambitious multimedia educational centre, we could still start grassroots education.

Papa wanted me to keep working on the new project, but this was not what I had signed up for. There was no detailed planning, no thought of stocking rates, and I had to ask myself some serious questions, the main one being what was I doing this for? My name in conservation circles was at stake.

I regretfully declined, and my adventure with Papa and the DRC drew to a close, although I had a feeling it was only temporary. As a conservationist, the mesmerising lure of Joseph Conrad's 'heart of darkness', as the DRC is erroneously known, is as addictive as a narcotic. The magical wildernesses, the surging rivers as beautiful as tropical beaches, the vibrant people and the enormous potential affected me profoundly.

It is true that I had not achieved what I had set out to do – in fact, nowhere near. My initial plans of importing rhino from South Africa for a breeding herd at N'Sele protected by army troops never materialised.

But it was more than that. Success in Africa is not always measured in material gain. I learned a life-changing lesson that anyone who wants to thrive in the wilds of Africa should experience. And there is no better place than the DRC to teach it to you.

It is this: every day that you are still standing, you are winning. It's like getting knocked down constantly in a boxing ring, dealing with corruption, language barriers, civil war, often horrific disregard for life . . . but somehow just getting up and being on your feet signals victory. That's why people such as me, whether we're called idealists or idiots, keep coming back. It's that simple, and that complicated.

As I flew out from Kinshasa after my Kimpoko visit and looked out of the plane window at the seemingly infinite jungle below, I knew that this was not over. Subliminally I sensed that I had unfinished business with the vast, chaotic, anarchic, beautiful land. The siren call of this enigmatic country would seduce me back.

I was not wrong.

CHAPTER NINETEEN:

Saddling Up for New Solutions

The Congo project had been a fascinating if sometimes crazy ride, but now it was back to grassroots conservation work at Project Rhino.

My priority was rolling out Rhino Art into other areas, and as some of the old faces had left I started recruiting a new squad of stalwarts.

We soon had a top-class team, and it's not just me saying that. In the line-up was David Pattle, a highly respected labour lawyer; Adine Roode of Hoedspruit Endangered Species Centre (HESC), who runs a world-class rhino orphanage; Ricki Kirschner, who is doing ground-breaking educational work in Timbavati, home of the famed white lions; and an outstanding young man called Divan Grobler, who was part of a team that courageously exposed canned lion hunting.

I first met Divan when I was a guest speaker at the World Travel Market in 2016. He is the rhino specialist at Aquila Private Game Reserve in Touws River, about two hours from Cape Town. He is probably more famous, however, for going undercover in the hard-hitting documentary *Blood Lions*, which showed hand-reared, caged lions being 'hunted' by unprincipled trophy seekers. The lions aren't tame – no lion really is – but they have no fear of humans so won't evade them. To call that hunting is like calling yourself an angler by throwing dynamite into the water.

With Divan is Hunter Mitchell, who at ten years old was named 'Visionary Wildlife Warrior for 2016' by the Australia Zoo Wildlife Warriors, an organisation founded by the late, great Steve Irwin. Hunter, from Somerset West, raised thousands through crowdfunding to finance the foster-rearing of an abandoned rhino calf called Osita at Aquila reserve. To enlist these conservation heroes into Rhino Art was a significant boost to the project.

Interestingly, the orphan Osita – which means 'From today onwards it will be better' – was shown how to graze by a goat, introduced both as a companion and tutor to keep human interaction to a minimum. Just another reason why I love goats so much.

Obviously, the expanding Project Rhino was gobbling up money. We needed to raise funds urgently. As a result, myself and professional cyclists Mark Carroll and Cliff Wills created the ultra-tough uBhejane Xtreme Mountain Bike Challenge.

I love cycling fundraisers as you have fun, get fit and do things for a good cause. It's the ultimate win-win, and I have been involved in several, including the Wiesenmaier sisters' BuyNoRhino cycling odyssey around Vietnam, and Wayne Bolton's epic 3500-mile ride through all nineteen of South Africa's national parks in aid of rhino awareness.

The uBhejane Xtreme, arguably the hardest time-restricted mountain bike challenge in the world, is raced from the town of Hillcrest west of Durban to Hilltop, the flagship camp in the Hluhluwe–iMfolozi Park. It is an event with attitude, made even tougher as the 210-mile route is contested in the height of the steaming KwaZulu-Natal's summer. Much of the course is studded with steep hills and the cut-off time is sixteen hours, which means legs are pumping non-stop, even downhill. Most important, it is a charity endurance event, with no winners – except rhinos.

The first uBhejane Xtreme was held in 2014, but was a low-key affair held purely to raise some much-needed money for Project Rhino (*uBhejane* is the Zulu word for black rhino).

We changed all that the following year, raising the profile and the stakes. It is now a popular annual event, with the final 22 miles raced in the heart of the Hluhluwe–iMfolozi Park where cyclists may be accosted by one or more of the Big Five. That's great for game viewing from the comfort of a car, but a little more edgy on a flimsy mountain bike and could result in a literal ride for your life.

Only in Africa could such an event be held. If anyone in Europe or America tried to organise a race with lions, elephants and rhino having ringside seats, the health and safety brigades would have a seizure just reviewing the application.

The formalities are simple. The only entry requirement is that contestants have to cover their own expenses and raise a minimum of R5000 (about £280) for rhino conservation.

The figures speak for themselves. In four years, close on a million rand (about £56,500) has been raised.

In the memorable 2017 race, the 'voice of cycling' journalist Phil Liggett, former South African rugby captain John Smit and Sibusiso Vilane, the first black African to summit Mount Everest from both north and south routes, were at the starting line. The gruelling event is now solidly on the cycling calendar, according contestants much prestige just to cross the finishing line.

We needed that money urgently. The rhino wars were revving at full throttle. In 2014 and 2015, poachers killed 2,390 rhinos. In other words, more than three animals were murdered each day of the year, for two years running. They were slaughtered for keratin, the same stuff in human fingernails. The bush was soaked with rhino blood.

Although the horn wars were raging almost everywhere in the rural parts of the country, KwaZulu-Natal was the spearhead. This was for two reasons. Firstly, the province has the greatest density of both black and white South African rhinos, and secondly, while poaching was endemic in the Kruger National Park, there was a noticeable shift southwards, particularly towards the Hluhluwe–iMfolozi Park. Poachers now considered KwaZulu-Natal reserves to be a softer target than the famous Kruger, which is far better funded with a lot more boots on the ground.

Another major problem facing the Hluhluwe–iMfolozi Park is that a tarred public road almost surgically dissects the centre of the reserve, providing easy access and escape paths for poaching gangs. The R618 is a major rural route to the market town of Mtubatuba and runs through the most vulnerable part of the reserve, known as the Corridor. Previously, the Corridor was fenced off when Hluhluwe and iMfolozi were separate parks. When the two amalgamated in 1989, the fences were taken down to provide free movement of animals.

Unfortunately, it also provided free movement of poachers.

There was no doubt that Ezemvelo KZN Wildlife had a Herculean problem on its hands. The sheer scary statistics tell the story. In 2014, the province lost 111 rhinos to poachers. The following year it increased marginally to 116, rocketing to 193 in 2016 and 222 in 2017 – almost one rhino killed every twelve hours in a single province. As there are fewer than 18,000 white rhinos and a mere 5000 black rhinos still roaming the planet, you don't have to be a mathematician to grasp that statistically this is a wildlife holocaust.

Indeed, it is not far-fetched to say the current rhino wars are comparable to a blistering-paced action novel. Except this is no figment of an author's fervid imagination. The fight to the death between rangers and wildlife slaughterers; the rivers of blood, both animal and human, are all terrifyingly real. The heroes are hard men not afraid to put their lives on the line to defend vulnerable wildlife. The heavily armed villains are ruthless killers. The backdrop is the unforgiving African bush, the wildest tracts of land in the world and sometimes so thick with thorn trees that rangers often only accost poachers when they stumble upon them. Firefights are short, sharp and vicious, and fought at close range. Or under a full moon when poachers traditionally strike as visibility is better.

The poachers are skilled trackers and bush survivors. It takes a gang a mere twenty minutes to immobilise a rhino, dehorn it and melt into the wild. Often the only sign that an animal has been mutilated is vultures circling in the thermals high above. Unless they are caught red-handed, or involved in a firefight against determined, well-trained men, the odds are in the poachers' favour.

Also, what the grim statistics do not tell is that a poached rhino sometimes has a calf with her. The baby is sometimes not killed as the poachers don't want to waste a bullet for a stub of keratin. Instead, they slash the infant with axes or machetes to drive it from its mother. The baby seldom survives, either dying from its wounds or thirst, or, without a mother to protect it, it's devoured by other predators.

Then, of course, many of the slaughtered rhino are pregnant. If this was added into the overall head count, the picture would be even bleaker.

There is also the human toll, and not only in lives. Rangers are usually tough, taciturn people, but the emotional cost is dreadful. Many are suffering from PTSD after continuous encounters with horribly maimed animals. The constant stress of living by the gun and seeing unimaginable brutality on a weekly, if not daily, basis would emotionally cripple most people. Rangers are no different – they are only human. Ezemvelo's legendary vet Dr Dave Cooper, a grizzled bush veteran who now spends at least a third of his working hours doing post mortems on slaughtered rhino, said in an interview that he has shed more tears in the last five years than the rest of his life combined.

There are even reports of poachers buying properties on the borders of reserves to have ringside seats, so to speak. Not only can they virtually track rhinos from their homes, they can also act quickly if they know a rhino is in the vicinity. It's difficult to get more brazen than that.

What makes it worse is that the authorities know who many of the poachers are, but are unable to arrest them due to lack of hard evidence. But even when there are arrests and poachers are brought to court, they are almost always given slap-on-the-wrist fines.

The most disturbing example of this was when the alleged rhino-poaching kingpin, Dumisani Gwala, was arrested in 2014. At the time it was hailed as the biggest bust ever as police said that 80 per cent of all rhino horn trafficked from KwaZulu-Natal allegedly went through Gwala to markets in the Far East. Gwala faces multiple wildlife charges plus one of

White rhino calf in the Hluhluwe-iMfolozi Park, Zululand. Most severely endangered rhinos slaughtered by poachers in the province of KwaZulu-Natal are from this world-class game reserve. *(© James Glancy of Veterans for Wildlife)*

Thandi the miracle rhino, left barely alive after being brutally dehorned by poachers with machetes. She was saved by a veterinarian team led by my brother William — the first rhino ever to survive a machete attack where poachers hacked the horn off at the roots. *(© Roger Paul Mills)*

William treating Thandi. At first, he wanted to euthanise her to end her excruciating agony. She not only survived but has now given birth to a calf of her own – a true wildlife miracle.
(© Roger Paul Mills)

Grant in his trademark Zulu *Umblaselo* trousers with *Sky News* presenter Emma Crosby after a live interview at the Sky News centre in London. (© Jennie Munro)

Grant with Zulu leader and ardent conservationist Prince Mangosuthu Buthelezi.
(© Chris Galliers)

Rhino Art action. The ground-breaking project, in which today's youth is given a voice to spread the conservation message, is set to reach a million children by 2021. *(© Grant Fowlds)*

Kingsley Holgate with a group of learners at Ashton International College, Ballito, after being handed a cheque by marketing manager Jeannie Habig. With him are musicians David Jenkins and Maqinga Radebe, who make musical magic at rhino events. *(© Grant Fowlds)*

During the civil war in Mozambique in the 1980s, Paul Dutton darted this, the last of the country's White Rhino, wounded by poachers in the Maputo Elephant Reserve. It later died of septicaemia from the bullet wound inflicted by the poachers. *(© Elisabeth Specht)*

Imfolozi sunset with wilderness legend Mdiceni Gumede and Grant. In the background is the Hluhluwe-Imfolozi Park, the scene of several brutal slaughters of rhino by poachers and where highly effective horse patrols are now in force. *(© Angela Fowlds)*

(above) Peter Eastwood, Grant and John Kahekwa tracking gorillas in the Kahuzi-Beige National Park, Democratic Republic of Congo. In the background is the magnificent silverback Bonané. *(© Marcio Lisa)*

(left) Monica the chimpanzee and Grant at the Parc de la N'Sele reserve, Democratic Republic of Congo. The baby chimp was an orphan and it was heartbreaking for Grant to have to put her back in a cage after visits.
(© James Wamboy)

Narrow escape: the wrecked Land Rover after the near-fatal accident on the N2 between Grahamstown and Amakhala in which Jess, Georgine and Alice Fowlds were seriously injured. *(© Tracy Weeks)*

Grant with daughters Georgina and Jess at Sunnyside (now Amakhala), in a field of wild veld flowers. *(© Angela Fowlds)*

Atherstone Nature Reserve elephant guards who took part in the capture of threatened elephants that were relocated to the Eastern Cape. *(© Grant Fowlds)*

A darted Atherstone Nature Reserve elephant being loaded onto a truck before the long journey to a new reserve and safety in the Eastern Cape. *(© Grant Fowlds)*

attempted murder as he allegedly tried to run over the policeman arresting him. He has pleaded not guilty to all counts.

The trial has now been postponed eighteen times due to a litany of excuses and apparently 'lost' dockets. There was even a call by the prosecutor for the removal of the presiding magistrate on suspicion of corruption. Instead, the prosecutor was removed.

But perhaps more alarming for those of us out in the field is that rangers and conservationists were called to give classified information in open court that basically provided poachers with a 'how to' guide to evading the authorities.

I had been closely following the Gwala case and was interviewed by *National Geographic* magazine, which thankfully managed to attract some international attention. But even with the glare of the global media, the apparent circus continues. After four years, Gwala is still out on bail and hearings are still being postponed.

Conservationists were fighting a losing battle every inch of the way. We had to find new methods to fight back and boost morale. We needed new tactics, new energy and new vision.

It was during this depressing period that I met a remarkable woman called Karrie Hovey, an American artist from northern California who has exhibited around the globe. She is a fiercely committed conservationist and gets her inspiration from the natural world, so we had an instant rapport.

In 2014, she and her husband Charles Merrill took part in the JoBerg2c, a nine-day, approximately 560-mile mountain-bike race from Johannesburg to the Indian Ocean. It raised a lot of money for conservation, some of it being channelled into Rhino Art.

Being an artist, the unusual concept of children drawing as a means of wildlife education captivated Karrie. She 'adopted' Rhino Art as a charity and started raising money for projects in eight schools in Zululand. It was a complete shock for her to discover that most of the rural pupils had never seen a live rhino.

Soon afterwards, I brought her out to KwaZulu-Natal on behalf of Project Rhino and introduced her to Dennis Kelly, section ranger of the Nqumeni area of the Hluhluwe–iMfolozi Park. Dennis is one of the most effective Anti-Poaching Unit (APU) operators in Ezemvelo, and he and Karrie got chatting. She asked him why horses were not used on patrols any more.

It was a good question, even though – ironically – Dennis *did* use horses. He was one of the last remaining Ezemvelo rangers to do so, and said they were invaluable assets.

Most of the horses still owned by Ezemvelo, however, were past their use-by date. Due to stringent budget cuts, the health and age of the remaining animals, crumbling stable structures and new technology, they had been discontinued.

Dennis was a big fan of horseback patrols, telling Karrie they were silent, could go where vehicles couldn't, wouldn't spook other animals and alert poachers, didn't need headlamps in the dark, could be refuelled with grass, and would give advance warnings if predators such as lions were around.

Horses also camouflage human tracks, allowing rangers to move undetected, and being on horseback they do not 'desensitise' rhinos to human presence. Rhino who are accustomed to humans will not flee from poachers.

The cherry on top was that no poacher could outrun them, and to face a horse in full gallop with flailing hooves … well, that alone was a significant deterrent.

In short, they were ideal.

So, persisted Karrie, if horses were so effective, why had they been discarded? Was the reason financial or bureaucratic?

Dennis shrugged. It was a bit of both, he said, but lack of finance was the main obstacle. Ezemvelo didn't have the funds to run stables, grooms, train riders, buy feed and do patrols in such a vast park. They also needed fresh blood. Their animals were not getting any younger.

'I'll get you horses,' said Karrie. 'I'll crowdfund – and guarantee $15,000 for you.'

Dennis and I whistled. That, with the exchange rate at the time, was more than R280,000. It would go a long way.

Part of the magic of working in conservation is thinking outside the box. That's exactly what we were doing and I loved it. When something doesn't work or is not proving effective, try something else. We were doing everything we could in the rhino wars, but still not winning. We had spotter planes in the air, boots on the ground, armed Anti-Poaching Units patrolling the bush, tracker dogs sniffing spoor, grassroots education programmes such as Rhino Art entrenched in schools, and we were getting good cooperation from community leaders. We seemed to have everything.

But we didn't have horses.

Now we did.

CHAPTER TWENTY:

Hooves for Horns

For those who love the outdoors, there is nothing more enthralling than trekking in the backcountry on horseback.

It's cemented into wilderness folklore, where horses were key for long-distance off-road transport and men and women of the bush often spent days camping out on the trail. We were going back to grassroots.

But perhaps just as entrancing was the 'romance' of the idea of reintroducing horses. We would be using animals to protect animals.

Horses have an exalted status in many parts of South Africa, right up there with the mustang in America. Legends abound of mystical relationships, and perhaps the most famous is that of farmer Wolraad Woltemade, who in 1773 rode his horse Vonk into pounding surf to rescue sailors from the galleon *De Jonge Thomas* wrecked off Cape Town. He made seven trips, pulling panic-stricken men to safety. Tragically, Woltemade and Vonk drowned on the eighth attempt as the ship broke up. Of the fifty-three survivors, fourteen owed their life to the heroic farmer and his horse. Vonk's body was never recovered.

Another epic ride was that of KwaZulu-Natal trader Dick King and his sixteen-year-old servant Ndongeni, who raced on horseback to get help when Boer commandos besieged the British garrison at Durban in 1842. The nearest military outpost was at Grahamstown 600 miles away. King was lying low on a ship in Durban's harbour at the time. When night fell he and Ndongeni swam their horses across the shark-infested channel. They then began a race for their lives, with King reaching Grahamstown in ten days on a trip that usually took seventeen. Ndongeni, who had no saddle or bridle, had to turn back after 250 miles – also an amazing feat by a youngster.

Alerted by King, a navy ship sped up the coast to Durban and the blockaded British soldiers, who were eating boot leather to stave off starvation, were saved. King was regarded as a hero by the British, but obviously not the Boers. However, no one disputes that his incredible ride through thick bush and fording more than a hundred crocodile-riddled rivers in ten days is an astonishing equestrian achievement. As Ndongeni could not swim, King had to pull the Zulu's horse across rivers with the youth still mounted.

Interestingly, King's horse, which was named Somerset, was originally stolen from the Boers by a Griqua bandit and sold to the British, which is how King got it. So the animal's bloodline would be that of the hardy Boerperd, the same as those we are now using on modern anti-poaching patrols.

Horses have also always been an essential part of the country's conservation story. Since the first game reserves were proclaimed more than a century ago, most rangers relied exclusively on horses to patrol the vast wild places where the only routes were animal tracks.

They were expert horsemen, and many of the older generation still are. They had to be. In Zululand rangers used horses to capture herds of antelope by the simple and extremely dangerous expedient of chasing the animals at full gallop into holding pens. Nowadays, this is done by helicopters, whose pilots are as skilled in the air as the horsemen were on the ground.

Perhaps the most celebrated conservation triumph using horses was Operation Rhino under Ian Player where, as mentioned earlier, white rhino on the verge of extinction were captured to repopulate their former ranges. Rangers tracked rhinos on horseback, darted them, then called in the transport trucks. Without horses, this would have been impossible in the thorny scrub and rugged veld.

This, for us, was a poignant history lesson. Horses were pivotal in saving rhinos the first time the species nearly went extinct. We were now calling upon them again.

It was not just a case of redeploying the animals Ezemvelo still had. Most of the horses had not been used for a decade. We needed fresh blood, skilled grooms, advanced equestrian training, as well as saddles, gloves, tons of horse feed, bulletproof vests and binoculars. With Karrie's crowdfunding contacts, this was going to happen.

Karrie created Project Thorn, an acronym for 'Tangible Help Our Rhino Needs', to formalise the goal of reviving horse patrols. We outlined the plan to Ezemvelo and they liked it. It wasn't rocket science as all rangers know that regular patrols are the most effective way of preventing poaching, and horses are the most effective way of patrolling isolated areas. It's the bush version of bobbies on the beat.

Project Thorn proposed using horses for three key tasks. Firstly, and most important, they would be deployed as law enforcement – a fancy phrase for catching bad guys; secondly, they would patrol fences to detect breaches in security perimeters; and finally, they would be used for veterinary and research work. The third task would be more scientific than security, giving researchers opportunities to get closer to animals in the wild.

It's difficult to overestimate how effective horses are in the wild. People walking through bush are almost always highly conspicuous to wildlife. Even those who think they are skilled outdoorsmen, moving stealthily and constantly checking wind direction, often stick out like drunks at an AA meeting. But put a human on a horse and the dynamics change beyond recognition. Horses are not predators. They do not alarm other animals. Consequently, not only do they blend in beautifully, they disguise both a human's distinctive outline and the distinguishing smell. A human on a horse is not regarded as an intruder by other wildlife. But a human on his own is considered an alpha predator.

It's not just horses. Some safari companies offer game viewing from the back of tame elephants. The effect on surrounding wildlife is exactly the same – total acceptance. A buffalo or white rhino usually won't bat an eyelid if a rider approaches, although the notoriously bad-tempered and short-sighted black rhino have occasionally charged horses, even goring them. But then, black rhinos have been known to charge elephants, which unsurprisingly does not work out well for them.

The only drawback is that lions regard horses as gourmet meals and a human rider can be collateral damage. At least one Hluhluwe–iMfolozi ranger has lost his life that way. He was pulled off his horse and eaten by lions in the Nqumeni section during the early 2000s, which put a bit of a damper on horse patrols for a while.

Even with lion, though, the odds are still tilted in a mounted ranger's favour and a horse can usually outrun a giant cat despite having a burly guy on its back. And that burly guy would without doubt be vigorously urging his steed to run even faster. Also, the mere whiff of a lion is enough to stop an experienced bush horse in its tracks. So rangers pay a lot of attention to a suspicious-looking thicket if their horses start getting edgy.

We now had the funds and the go-ahead from the authorities. It was all in black and white, with Ezemvelo, Project Thorn, the conservation charity 12hours (that's how often a rhino is killed) and Project Rhino signing a service-level contract with myself as facilitator. Central to the agreement was a reciprocal confidentiality clause, essential when dealing with rhino – if sensitive information regarding their movements is leaked, it's tantamount to an animal's death sentence. No game reserve anywhere in Africa ever mentions a rhino sighting over a radio. It's a rule carved in granite.

Then, to my surprise and delight, I discovered that horse patrols were also about to be introduced in the Eastern Cape. One of my brother William's vets, Dr Danielle Jackson, was a horse fanatic and, completely independently of us in KwaZulu-Natal, decided to start a unit called Equine Anti-Poaching, based at Amakhala.

Danielle joined William's team in 2014 after qualifying as a veterinary surgeon in the UK and fell instantly in love with Africa. She was soon in the thick of the poaching wars, and on one occasion was instrumental in tracking down a baby rhino whose mother had been killed, saving the infant from an awful death.

Although Dani started the Eastern Cape project after us, she had anti-poaching patrols up and running under the umbrella of the charity African Rhino Conservation Collaboration (ARCC) long before we did. The key factor was that she was dealing with the private sector, while we were bogged in the quicksand of bureaucracy and red tape. The American-based non-profit Global Conservation Force provided the funds and Dani was able to buy horses and equipment and had the unit out in the bush while we were still organising meetings about meetings.

To expedite anything with parastatals is a political minefield, but I believe that is true in most countries. It was no different for us. We were hampered by endless delays, legions of lawyers, forests of paperwork, mountainous egos and, if one aggrieved person felt they had been bypassed, everything got kicked back to the drawing board. If we at Project Thorn didn't have skins as thick as the rhinos we were trying to protect, the concept would have bombed.

The most rundown stables were at a section called Mbhuzane, ironically the area that needed patrols the most. More rhino had been poached there than almost anywhere else on the reserve.

So that was a good place to start. Except no one apart from Project Thorn seemed to think so. We had no support whatsoever in getting the stables up to required standard. We were promised that basic ingredients such as cement and poles would be supplied from Ezemvelo stock, and that honorary rangers – there are about eight hundred of them – would help.

No poles nor honorary rangers materialised.

Just as I was yanking my hair out in tufts, by a stroke of amazing good fortune I bumped into a New Zealand round-the-world yachtsman moored at Richards Bay. It happened, as many great things in life seem to do, over a frosted beer. Mark Lumsden, his partner Kim Bernard and I were quaffing something long, cold and foamy at the yacht club when Mark asked what I did for a living. I said I worked with rhino.

That got their attention as Mark and Kim are passionate conservationists. Not only that, Mark is a top-notch engineer and carpenter. When I told them the hassles I was having just getting bog-standard infrastructure repaired for rhino management, he and Kim offered to lend a hand.

I gave Mark and Kim five days to fix the stables at Mbhuzane. They took four, re-roofing and replacing rotten, ant-eaten poles and planks so expertly that it was as good as new. I was in the DRC at the time and got regular photo updates on my cellphone. I was astounded at their spectacular progress. What they did in those few days would have taken almost anyone else a month or two.

It was now all systems go. We were ready for the horses.

Unfortunately, that was not how the section ranger of Mbhuzane saw it. He accused me of coming to his area without permission and bringing in 'unauthorised people', a highly inflammatory accusation in the context of the rhino wars. When I pointed out that the 'unauthorised' visitors, Mark and Kim, were actually unpaid Good Samaritans, he told me I was no longer welcome in his section. This was a massive shock as initially he had been one of our most avid supporters.

The worst was when the horses arrived from the Griqualand capital Kokstad and the same section ranger refused to allow them onto the reserve, saying head office had failed to 'communicate properly' with him. Accompanying the horses was our newly trained and keen-as-mustard groom, Dennis Gumede. Dennis is the son of Ezemvelo's wilderness legend and one of my wildlife mentors, Mdiceni Gumede. Dennis called me to try and sort the mess out.

There was an unpleasant standoff at the gate. Everything we had worked for was in jeopardy because a section ranger claimed he had not been kept in the loop.

Despite my intense annoyance, I had some sympathy for the ranger as he had an almost impossible task and was without question at the coalface of the poaching wars. He was a brave and dedicated man. The fact that he had just lost some rhinos was a crippling blow, and morale in his section had plummeted to an all-time low. So I understood where he was coming from. But picking on people trying to help him wasn't productive.

We stood arguing in the sun for several hours. Eventually he agreed to open the gate and let the horses, still cooped in transport trucks, into the stables with the proviso that he was absolved of all responsibility. He said budget cuts meant he 'could not babysit horses' as well. I told him we had already raised the necessary funds, and pointed out we had a contract signed by Ezemvelo management. That didn't seem to impress him much.

But at least the horses were in their new home, where they adjusted well in spite of the arid conditions as we were in the middle of a drought. The animals were definitely up for the task.

The only problem was that they were not being used for that task. For five months they did not go on a single anti-poaching patrol – the key reason for their existence. They languished in their stables, being used for virtually nothing apart from some ecological monitoring. I shook my head, ruefully considering how impressively Dani was doing running horses at Amakhala compared to the bureaucratic cluster-flap we faced.

Even more frustrating was that while the horses were regarded as unwelcome intruders, I still had to sort out all running and administrative problems. I was called out to the stables at Mbhuzane multiple times, doing a 250-mile round trip from my house for minor issues that could have been solved in minutes by someone on site. To add insult to injury, I also had to beg permission to enter in advance or risk not being let in through the gates.

At the same time, I was pestered by donors for progress reports. This was logical as they wanted to see their money in action. The only problem was that no progress was being made. All I could do was say as ambiguously as possible that things would be happening 'soon'.

Then, just as abject despair bordering on depression set in, everything suddenly fell perfectly into place – as so often happens in conservation. In this case, it was when Ryan O'Connor from Ezemvelo's Wildlife Crime Department and District Conservation Manager Des Archer, a former Bush War veteran and experienced horseman, arrived at Mbhuzane for an inspection.

Des and Ryan started asking probing questions. Such as, why was the rhino war going so badly? What were managers doing on the ground?

Of particular concern was that rhino poaching was intensifying beyond belief. It was no longer only spiking during full moons. Rhinos were now also dying on pitch-black moon-less nights, regardless of weather. This had the tragic side-effect of an escalating number of already dehorned animals being killed. Poachers would normally leave those animals alone, rather than waste effort and bullets for little blood-money payout. But now killers hunting on black nights without the visual benefit of a full moon would not know a rhino had been dehorned until they shot it.

While all this officious wrangling was going on, Dennis Kelly continued running horse patrols in Nqumeni, the only section ranger regularly to do so. As a result, few rhinos were poached in his section. Dennis said straight out that was mainly due to his horses. He actually wanted more as one had been killed by lions when it wandered off looking for greener grazing. Whenever he or his staff heard big cats roaring at night – Africa's purest primal sound – their first priority was now to rush to the stables to check if the horses were safe.

Dennis's success and commitment persuaded Ryan and Des to mandate that horse patrols be introduced to all sections of the reserve as soon as possible. All head rangers had to comply.

Soon afterwards, Dennis transferred to Makhamisa, the oldest, southernmost section of the Hluhluwe–iMfolozi Park where the main routes are ancient animal tracks.

The new section chief at Nqumeni was Sibonelo Zulu, whose right-hand man Corporal Pumlani Dlamini was a scarred survivor of the poaching wars, although not in a rhino context. Dlamini had been involved in more Wild West-style shootouts than Wyatt Earp and Wild Bill Hickok combined against cattle rustlers from Lesotho on the southern KwaZulu-Natal border. Lesotho is a tiny kingdom surrounded by the jagged Drakensberg and Maloti mountains, and there are regular border skirmishes between Basotho bandits and Ezemvelo rangers. The Basotho are among the hardiest mountain people in the world. They can clamber up a sheer cliff face with a gutted sheep carcass on their back faster than most people can run. As Sibonelo said, Corporal Dlamini was a good man to have on your side in a fight.

Sibonelo was a godsend. One of the first things he did was ask me to get him more horses. He was not just a breath of fresh air – he was a hurricane. I shook his hand vigorously and promised him three more steeds.

Included in the new batch was a beautiful Appaloosa called Orlando and an Arabian crossbreed, although most of the horses were Boerperds, just as Dick King had used. This is a specific mixture of bloodlines blended in South Africa, basically a fusion of genes from powerful old Cape farm horses and the now extinct but legendary Boer horse, used by Afrikaner commandos in the Anglo-Boer war. They are exceptionally hardy and adapt perfectly to the harsh wilds of the iMfolozi valley.

Included in the training was teaching our horses not to panic and shy away from lions. That's like training someone to remain calm when approached by an insurgent wearing a bomb vest. This was done by Californian former champion jockey, Roxane Losey, a director of the Global Conservation Force, which has been pivotal in providing patrol equipment, including bulletproof vests. She trains the animals at Amakhala, gradually bringing them closer and closer to lions, cheetah and rhinos, so if something sudden happens they don't stampede wildly. The danger with horses in lion country is almost exclusively limited to summer when the big cats hide downwind and ambush in head-high elephant grass. Very rarely will a lion catch a horse in winter when the grass dies.

Sibonelo brought some fundamentally fresh thinking to the project, initiated by Dennis before him. Horsemen in his section were now patrolling up to 25 miles a day over all types of terrain. They were also on standby as a reaction force at night, waiting and watching from hilltop lookouts, listening for gunfire, searching for flickers of flashlights. As horses can gallop at 30 miles an hour, their reaction time in the unforgiving bush is phenomenal. A human, at best, can average about 6 miles an hour in those conditions.

Nqumeni currently has seven animals on daily patrols. All rangers on horse duty have GPS CyberTrackers in their kit so donors can see what is happening in real time and can judge for themselves how successfully their money is being spent. That is far more effective than dry, emailed progress reports.

But the most exciting statistic is that, at the time of writing, the Nqumeni section has had 207 days in a row with only one rhino poached. That's still one rhino too many, but the fact that this dovetails with the introduction of horse patrols is more than coincidental.

Perhaps 207 days, little more than six months, may not sound like a long time to people sitting in air-conditioned suburban homes or high-rise offices.

But out in the white-hot crucible of the rhino wars, it's a lifetime.

CHAPTER TWENTY-ONE:

Riding for Rhinos

Having waded through mountains of paperwork and talked myself hoarse in putting the case forward for equine patrols, I decided to actually go on one.

To test the programme 'in anger', so to speak.

I grew up with horses, but wouldn't say I'm an expert. I first saddled up at the age of six on Cheeka, my Shetland pony, who made a sloth look enthusiastic. He only operated on sugar cubes and carrots and when the groom saddled him, he sneakily blew out his stomach to stop the girth biting into his substantial gut. After a few yards, he passed wind mightily and his belly deflated like a balloon, loosening the girth and causing the saddle to slip. The rider, in other words me, usually followed suit.

Often there were tears, and Cheeka mastered stalling tactics to such an extent that the most successful rides were seldom more than a couple of hundred yards.

My next horse was Lady, a brown mare who was as sneaky as Cheeka. While she didn't extend her belly during saddle-up, she instead walked so slowly that I could have got to wherever I was going faster under my own steam.

But as soon as I turned for home, she would erupt into an explosive gallop that rivalled an Aintree steeplechaser. This at least taught me to cling on for dear life when she sniffed the home straight.

Then there was Shannon, a snow-white gelding with pink eyes, who was partially blind due to albinism. Riding into thorn trees was a common occurrence. Tolly, my loyal goatherd, had more faith in horses than me and his stallion Duke towed his Scotch cart around the lands when tending animals.

Perhaps I could be forgiven for ditching horses for more reliable transport with the onset of teenage adolescence and interest in girls. My first mechanised transport was a Honda monkey bike, so called because the rider dwarfs the tiny machine and looks somewhat simian in the saddle. However, it was ideal for ferrying girlfriends to the farm for weekends, which I like to think meant I didn't overly resemble a monkey. I must mention that even though girls were suddenly of interest, they never displaced my goats.

So who would have thought four decades later I would be lobbying for horses to fight in the rhino wars?

The more pressing question was this: could I bluff the crack teams at the frontline that I was as adept at riding horses as I hoped I was in organising them?

I was soon to find out. My chance came with a programme that we helped on the Tugela Private Game Reserve near Colenso in northern KwaZulu-Natal. The reserve is a community-based project managed by Ant Arde, one of my wildlife gurus and himself an expert horseman. When Ant heard I was in his area, he invited me and Roxane Losey to come on a patrol.

The Tugela reserve had joined forces with Project Rhino

after experiencing horrific poaching problems for several years. Veterinarians had also recently performed miracle bush surgery there, rescuing a mutilated dehorned rhino, similar to what William and his team had done at the Kariega reserve.

I followed the progress of the reserve with interest as I believe community parks will be crucial to the future of conservation. Basically, these are rural lands expropriated from white farmers and held in a trust for the benefit of the original inhabitants. Unfortunately, a lot of these reserves have failed because the communities, most of whom are subsistence farmers, have no experience in wildlife management. But more galling was that some of the 'experts' appointed as mentors callously took advantage of this naivety, robbing trust funds of vital resources. In the case of the Tugela reserve, a man entrusted with the project ran his own shooting safaris instead of helping the community. It was only now with experienced and dedicated people like Ant in charge that the tribespeople – the actual owners – started reaping rewards.

I became involved with several community parks as a spinoff from doing work with Rhino Art. Thankfully, much of the expropriated land was too feral for farming, so wildlife tourism was the most cost-efficient economic route. We were creating game sanctuaries as well as kick-starting cottage industries for local people who now had a tangible stake and – equally important – say in their future.

Roxane was out from California bringing plenty of patrol kit for our stables in the Nqumeni section of the Hluhluwe–iMfolozi Park and like me wanted to experience an anti-poaching patrol first hand. She jumped at the chance to do some frontline duty in a saddle.

Apart from rhino poaching, hunting antelope with dogs on

Tugela was also rife and Ant's dozen patrol horses were proving invaluable in curbing this. Several high-profile arrests helped spread the word that galloping *amahashi* – horses – were now chasing down all poachers. This undoubtedly was a deterrent to the far more organised rhino killers as well.

Included in our patrol was Ant's son Alex who, like his dad, is as at home on a horse's back as a seal is in the ocean. With Roxane, a top jockey who has won 300 races in the USA, also saddling up, to say I was the weak link in the group was being polite.

We rode through beautiful savannah studded with umbrella thorns, abundant herds of buck and green thickets of woodland. It reminded me of the movie *I Dreamed of Africa* and I half expected Kim Basinger to emerge like an apparition from the long elephant grass. Or perhaps get a glimpse of Meryl Streep and Robert Redford sipping champagne while picnicking in *Out of Africa*. It was that surreal.

We crossed streams and climbed kopjes, stunning wild country, and for Roxane, who had been in the concrete heart of Los Angeles thirty-six hours earlier, it was like landing in another galaxy. Her jetlag evaporated like a fog. She was absolutely in her element.

Despite being the worst rider in the group, I still had to take media photos of a modern anti-poaching patrol in action for the press. This was not easy as I was doing all I could just to keep pace with my ace companions, and usually only caught up if someone stopped to relieve themselves in the bush.

At one 'pit stop', the surrounding elephant grass was an impressive sight, swaying in the breeze high above our heads even though we were on horseback. I pulled out my cellphone to take a video.

At that exact moment, as I sat on my horse with my cellphone in my hands and glasses perched on the edge of my nose, Ant and Rox decided to go for a little recreational gallop. Which was fine, except my horse, an Appaloosa, joined in the fun.

Off the animals sped, charging into the bush. I had both hands on my cellphone, which is essentially my mobile office. I can't function effectively without it. All my conservation and marketing contacts – some five thousand of them – are stored in the gadget's memory. If I lost that, it would be tantamount to erasing an entire filing system. So no matter what happened, I could not let go.

This meant that the reins were flapping around the horse's neck instead of being in my hands, and there was no way of putting the brakes on the stampeding beast. All I could do was shout at it in Zulu, which I assumed would be the orders with which it was most familiar.

The Zulu command for stop is '*wow*', which obviously means something vastly different in English. Unfortunately, my Appaloosa preferred English and decided that me shouting 'WOW!' at the top of my lungs was more a yell of encouragement than a plea to cease and desist.

No one, least of all me, knows how I remained in the saddle, albeit at extremely precarious angles. If I had tumbled, I could have been seriously injured on the rocky ground. Maybe even fatally at the speed my horse was galloping.

More miraculously, I managed to hang onto my cellphone. But I lost everything else, including my sense of humour. Ant and Rox, meanwhile, were laughing so hard they were crying.

To make matters worse, the video I shot for the media was

a blurred chaotic scramble. Totally unusable for anything except proof of my indignity.

The sight of the reserve's lodge, signalling the end of the patrol, was the most welcome spectacle in the world. However, even though I had shown I was not the most competent rider, I had unwillingly proved I could stay on a horse's back in an emergency. For that alone, I think I earned my spurs, so to speak.

Somewhat chastened but extremely happy with the way the horse project was going, I returned to another battlefield, this time on the far northern front, where things were not 'all quiet' by any means.

It happened when I received a newsletter called *The Conservation Imperative* – required reading in the conservation world and compiled by a friend, Bugs van Heerden. The story that caught my eye in this particular edition was that elephants in the Limpopo province's Atherstone Collaborative Nature Reserve were becoming a serious problem and up to eighty might have to be moved. If new homes could not be found, other less palatable options would have to be adopted.

This interested me intensely. In the most extreme cases, the phrase 'other options' sometimes means culling. To me that was unacceptable. Wildlife range expansion through community engagement, as we were doing with the Tugela reserve, was one of my key missions. I don't want any animal culled on my watch because we couldn't find it a new home.

I contacted the Limpopo authorities, who confirmed that the herd was expanding too rapidly, knocking down groves of marula trees that are prime nesting sites for endangered vultures, and generally upsetting the eco-balance. They now had 130 elephants on the land. They needed to get rid of more than half – eighty in total.

That's a lot of elephants. Too many for me as a one-off project. Juggling figures off the top of my head, I told them I could probably find new homes for about thirty-five of them reasonably quickly, and then raise funds to relocate the next thirty-five, while getting an international translocation permit for the final ten. The figures weren't set in stone, but at least the authorities knew that I was out there doing the groundwork.

Limpopo's conservation authorities said that although a cull had been tentatively approved in an Elephant Management Plan released more than a decade ago, it was an absolute last resort. In other words, a disaster-case scenario that was extremely unlikely to happen. This went a considerable way to allaying my fears. However, the mere fact that a cull was possible, even if unlikely, set off warning bells in my head. It showed, at the very least, that the situation needed to be handled with extreme delicacy.

The 23,500-hectare Atherstone reserve is north of Dwaalboom, close to the Botswana border. It consists mainly of vast savannah plains with bushveld and Kalahari grasslands, and apart from elephants is home to a significant population of black rhino, which was another reason why I was interested. As far as Rhino Art was concerned, any reserve harbouring the iconic animals was right up there on our radar.

The former owner, Norman Edward Atherstone, was originally a cattle farmer who re-wilded his lands like we had done at Amakhala to become the area's first game farmer.

A confirmed bachelor, he had no heirs and bequeathed his property to SANParks (South African National Parks) when he died.

In 1990, the Atherstone Nature Reserve was founded and

four years later it was renamed the Atherstone Collaborative Nature Reserve, also incorporating other private farms.

It's magnificent but very flat, unlike the hilly reserves of KwaZulu-Natal or Eastern Cape that I am used to. There are no peaks worth speaking of, and consequently no vantage points. If someone wants to scan the horizon, they have to climb one of the beautiful marula trees. The birdlife is incredible and, apart from spectacular vulture nests, there are at least seven species of eagle.

However, few people have heard of the park, and to have such a quintessentially British name in an Afrikaans and Sesotho-speaking stronghold – Atherstone sounds more like an English cricketer than wild bushveld – is in itself unusual.

To see what I was getting into, I decided to visit the reserve and took Angela and our daughter Alice with me. We were generously loaned a 4 x 4 Safari vehicle by Martin Steer, CEO of a company that supported our rhino and elephant programmes – otherwise the trip would not have been possible.

To my surprise Atherstone was closed to the public, although as conservationists we were allowed in. The manager – who by a stroke of good fortune came from the Addo Elephant Park next to Amakhala – told me this was due to lack of funds. They hadn't had a guest for three years.

Otherwise, it is a functioning reserve. It has its own graders and trucks, and dedicated rangers still patrol the borders, protecting the endangered black rhino. They are doing good work; two of the most notorious poachers in the north-west had their hands cuffed at Atherstone.

As some of the park incorporates private land, limited sustainable hunting is allowed and, like it or not, it is funds from those guys which keep the reserve viable.

Seeing first-hand the eco-damage from having too many elephants on land that couldn't sustain them, I decided we had to move fast. The animals needed new homes as a matter of critical urgency. I had to start moving right away.

I also discovered something else. At Atherstone, a new word was about to enter my vocabulary, a concept I had never heard of before.

Eco-scamming.

CHAPTER TWENTY-TWO:

Was It Eco-scamming?

American activist Phillip Hathaway first contacted me when I was working for Papa Kadima in the DRC. It was an email out of the blue and destined to cause me a lot of hassle later on.

He explained that he ran a fundraising organisation called Elephant Rescue (ER), and wanted me to join his board of advisors.

He said his board consisted of specialists from around the world and included top conservationists, expert safari guides, attorneys, directors of research institutes and even a senior cleric. I would be in venerable company. Also, his website claimed to have an impressive interactive audience of many thousands of conservation-minded followers.

It seemed legitimate, and I thought it would be a good idea to accept as it could be useful in nurturing my own fundraising profile in America. That's where most of the fat-figure international conservation money comes from. American backing usually results in a lot more publicity and cash, which is vital in the battle for the planet. So I accepted, and my name went up on the ER website.

I was primarily a rhino man, but also extremely interested in relocating surplus elephants as I had been part of the group that moved our first herd to Amakhala from Phinda nearly a decade and a half ago. This was the main reason why the story of the Atherstone elephants in distress caught my eye and prompted my visit to the reserve.

I contacted Hathaway, as I thought Elephant Rescue would benefit from the relocation campaign while I might get funding via ER donors in America. He said he would follow it up on his website.

I then met with the Limpopo conservation authorities, who outlined the habitat problems, all the while adamantly denying a cull was being considered. They showed me their Elephant Management Plan, which clearly stipulated that other options such as contraception, range manipulation and translocation would be tried first.

In other words, culling wasn't going to happen, despite the brief reference to it in the management plan.

I relayed this information to Hathaway, saying no cull was on the cards and that I was involved in finding the elephants new homes.

I thought he understood, but apparently not. Despite my assurances to the contrary, he latched onto the obscure reference to culling in the Elephant Management Plan. He then followed this up with an international plea on the for-profit petition website www.change.org. blowing it out of all proportion.

When I first saw the petition, I got the shock of my life. The headline read: 'Eighty elephants will be shot dead unless we move them right away. Please help.'

It got worse. 'ElephantRescue.net is petitioning Johan Kruger, who oversees the Atherstone Nature Reserve in South

Africa, to not kill the 80 elephants who live there,' was the first paragraph.

The petition concluded: 'This is an urgent matter of life and death so we need your signatures asking Johan Kruger not to kill these elephants. And, please post this petition on Facebook, Twitter and tell all your friends, relatives, co-workers and neighbours.'

Johan Kruger was the Deputy Director of Biodiversity for the Limpopo Department of Economic Development, Environment and Tourism (LEDET). To address an inflammatory petition directly at him, accusing him of planning a mass slaughter of elephants, was advocacy journalism at its worst.

Hathaway also claimed on his website to have found a home for the allegedly doomed elephants, 'safe from poachers and hunters', as it was protected by the army. That was certainly news to me. No one from Limpopo conservation or the Atherstone reserve knew anything about it. And these were their elephants.

He further claimed that an expert in translocating elephants was advising his network and once they had raised sufficient funds, the elephants could be moved. The figure of $150,000 was mentioned. To achieve this, supporters were being asked to donate money via a website link.

Apart from anything else, the highly charged wording seemed to imply to a global audience that callous conservation authorities in South Africa were hellbent on slaughtering animals. This was so far from fact it was laughable.

Indeed, Hathaway's plea on his own website took heart-tugging purple prose to new heights. 'You can make donations of any amount. The elephants thank you with all their heart.

We can't let these beautiful elephants die! This herd of 80 elephants is a close-knit, loving family and must be saved. With your help we'll save the fathers, mothers and babies.'

Talk about tossing a flaring match into a vat of gasoline.

I was now seriously alarmed, not to mention dismayed. This type of histrionic language would make kind-hearted animal lovers throughout the world go ballistic, and an avalanche of misplaced goodwill could jeopardise both the elephants and the Atherstone reserve. The elephants had to be relocated; there was no question about that. But well-meaning public outrage could thwart a calm, dispassionate debate and planning, resulting in the animals being forced to remain at Atherstone and further degrading the habitat while white-hot emotions erupted.

Equally alarming, as far as I was concerned, Hathaway had betrayed my trust. I had relayed my information to him about relocating the animals in confidence and was repaid with a provocative petition claiming a possible mass slaughter – which we all knew had been categorically denied.

I quickly banged off another email. 'Phillip, this is extremely concerning,' I wrote. 'We are not meant to be saying a word until we have a MOU in place. This news may halt the whole process.

'The Government of Limpopo and the park manager may also deny me access. I'm due there on Monday.

'What makes this worse is the park [Atherstone] is mentioned which is completely confidential.'

Unfortunately, I abbreviated the phrase Memorandum of Understanding to MOU, which I assumed he would be familiar with as it's a fairly common acronym in drawing up agreements. Hathaway pounced on this, saying he had no idea what MOU stood for or that my information to him was confidential.

Once again, the email ether sizzled. I replied: 'MOU is a legal Memorandum of Understanding. If we want the elephants to be saved, I believe the best way is to take the posting [of the petition] off the media immediately.

'The approach by handing a petition to Mr Kruger will burn the bridges before the project starts.'

Hathaway denied that either Atherstone or Johan Kruger had been mentioned in the petition and sent me the link to prove this. I clicked on the link. It took me directly to the change.org website and Mr Kruger's name was in the first paragraph.

Everything was starting to unravel. Even the influential Dr Marion Garaï, who chairs South Africa's Elephant Specialist Advisory Group, was dragged into the fray, confirming in a magazine article that while there was elephant overcrowding at Atherstone, 'culling 80 is not on the cards'.

'If at all possible, they (the authorities) would go for translocation as the first option,' she said. 'When change.org says the cull is scheduled for 7th September, that's total bull.'

You don't get much clearer than that. But the change.org petition was not taken down.

While all this was raging, the habitat situation at Atherstone deteriorated rapidly with the elephants doing long-term damage. The beautiful marula trees that had so entranced me when I first went to the park with Angela and Alice were now little more than withered stumps. This was not the animals' fault. They were doing what elephants do – pulling down trees to eat leaves or get at roots. This is not a problem in vast wildernesses, and actually generates new growth. But the Atherstone reserve could not support such a large herd. It was a factually simple, yet emotionally fraught, situation.

Apart from denying he had mentioned names when he clearly had, several other aspects of Hathaway's email intrigued me. Such as, 'I know Mr Kruger wants to avoid bad publicity. But this is rather silly. How can one move 80 elephants and keep it a secret?'

There was no mention of keeping it a secret. It was confidential, yes, but all management discussions in committee are confidential. Once a decision was made, it would be in the public domain. In fact, the final decision would be gazetted, as the Limpopo provincial government was legally obliged to do.

Hathaway also accused the Limpopo conservation authority of 'not thinking ahead', so any public relations disaster resulting from adverse publicity would be its fault, not his. In other words, he considered his inflammatory statements not to be his problem.

Finally – and most interesting to me – he said that the change.org appeal was more than a petition. 'It is also a way to raise funds for this project. We may, in fact, be able to raise one hundred per cent of the funds there.'

Ah. Was that the underlying motive?

This also begged another question: who would manage the funds?

I think we can be forgiven for suspecting what the answer to that was. There was no suggestion that anything illegal was happening, but there were certainly circumstantial concerns about transparency with people being asked to click a 'donate now' button in response to a highly emotive appeal.

Just as I was getting increasingly uneasy about Hathaway's conservation credentials, an article by veteran wildlife journalist Don Pinnock appeared, causing disquiet in nature circles. Writing in the *Southern and East African Tourism Update*, he said: 'While many good people and fine organizations raise public funds to support conservation, there are others who use conservation issues to raise money for themselves. The sad-eyed lion cub, baby chimps in a cage, adopting a motherless rhino or the urgent need to save creatures from culling are among many hooks to reel in sympathy cash from the web-linked public.'

The article was titled 'Eco-scamming: Making a killing on kindness'.

Equally alarming was an article published in Tulsa, Hathaway's hometown in Oklahoma. We were now in our man's own backyard.

Interestingly, the story was not from a conservation magazine. It came from the heart of academia – a publication called the *Collegian*, which bills itself as being the 'Proud newspaper of the University of Tulsa'.

Among other incriminatory information, which included the Atherstone debacle, was that the magazine cast some doubt on Hathaway's motivation for getting involved in elephant rescue. Hathaway said his passion for pachyderms was sparked by a report of a herd of more than three hundred elephants being ambushed by poachers on horseback wielding machine guns and hand grenades.

'All of them – even the little babies – were unmercifully slaughtered.'

Hathaway said this led him to form his own elephant rescue network as all other organisations were 'not excepting [accepting] volunteers'.

That is unlikely. Few, if any, conservation organisations, whether they are charities or for-profit ventures, turn away volunteers. These stalwarts are the backbone of the benevolent world, providing unpaid labour, and are always passionate about their causes. That's why they freely donate their time and skills.

But for me the real game changer was that according to the *Collegian*, Hathaway himself was operating under a pseudonym. His real name is John Stephen Mauldin.

So . . . why would anyone start a rescue organisation under a false name? Maybe it was because they were so modest they wanted their good deeds to go unheralded? I don't know.

While a pseudonym fundraiser is not illegal, it is highly unusual. But it begs another question: who would be willing to donate money to someone who doesn't provide his real name?

These two articles were dynamite in conservation circles. Soon afterwards, www.change.org posted a warning on Hathaway's online petition stating, 'Change.org has received flags that the facts in this petition may be contested. You should consider researching this issue before signing.'

The final straw was when Amakhala was mentioned on Hathaway's website. This concerned another appeal for funds to relocate a forest elephant named Can, apparently starving in a zoo in Abidjan, the capital of Côte d'Ivoire in West Africa.

Hathaway said he planned to relocate Can to Botswana, 'the only place in Africa where I know she'll be safe'. All that he needed was $90,000 to do so.

When information regarding Can's welfare was contested by the zoo, Hathaway implied that funds were instead being raised to relocate the elephant to Amakhala.

Amakhala? My home turf?

This was the first we had heard of this. It was absurd, not to mention bizarre. There was not a dry twig in a bushfire's chance that Amakhala would take a forest elephant from a West African zoo to live with wild savannah elephants thousands of miles from its natural habitat. To consider doing that showed how little knowledge of wildlife the self-designated 'elephant rescuer' had.

But even if Can had been a savannah elephant, Amakhala does not accept caged or captive animals. It would be inhumane. They would not survive in the Eastern Cape wilds.

I read all this, eyes growing wider by the second with amazement. The gargantuan lies – well-meaning or not – being told under the guise of saving photogenic animals were astonishing.

The apparent deception went deeper. According to the European Elephant Group, an organisation that improves the lives of captive elephants, Can was in fact being properly cared for. Funds had been raised to help her way back in 2012 and an Austrian elephant expert, Ingo Schmidinger, had already nurtured her back to health. So Hathaway's online appeal for financial assistance to move her to Amakhala – without even asking us – was several years out of date.

I decided I had to withdraw from this sorry scenario fast. The whole setup, in my opinion, stank to the skies. While there was no suggestion of anything illegal, I just wanted out. I told Hathaway to remove my name from his board of advisors with immediate effect.

He agreed that my profile was 'not a good fit' for his organisation. I could not have put it better myself.

I haven't heard from him since.

With Hathaway out of our hair, we could now concentrate on the real issue: finding new homes for a hell of a lot of elephants. It is an ongoing story, to which I will return later, and the complexities of moving eighty animals each weighing up to 6000 kilograms – or 6 metric tons – are challenging.

However, without relying on sensationalised headlines, impassioned social media pleas, for-profit petition websites and information that was often at best borderline, we started making good progress. The outstanding organisation ERP (Elephants, Rhinos & People) got involved, pledging 2 million rand, which is a lot better than any emotive commercial petition achieved. Elephant relocation is a slow process and all boxes have to be ticked before anything can be set in motion. ERP are experienced relocators, recently moving fifty-three elephants from the Mkuze and Pongola reserves in KwaZulu-Natal to Zinave in Mozambique.

I was somewhat disheartened by my experiences with so-called fundraisers after the Hathaway episode. I am not naive, but I did expect a little more honour, dignity and credibility among people who profess to be eco-warriors.

I soon found it. Not long afterwards I met a conservationist who is the complete antithesis of any alleged eco-scammer. A man of immense courage, unfailing optimism and granite stoicism, who with no funds, no government backing and scant international support does more for the planet than most of the rest of us combined.

He comes from the far east of the Democratic Republic of the Congo, the wildest stretch of arguably Africa's wildest land. A land I now knew well.

I like to think my meeting with John Kahekwa was fate. It was meant to happen.

CHAPTER TWENTY-THREE:

Humble in the Jungle

For most of the past two decades, my primary interest has been rhino conservation.

Not only are they among the most endangered animals in the world, I love everything about them; their character, their courage, their indomitable spirit and their dignity.

Yet my next venture took me down a road I had never even vaguely considered travelling. The great apes of West and Central Africa.

Previously, primates had not been high on my agenda, mainly because the three species in South Africa, vervets, samangos and baboons, are not endangered. They are flourishing. Vervet monkeys are so numerous that they are considered pests in some suburbs bordering the bush or beach dunes.

They also are not apes. Apes are usually not only bigger than monkeys, they do not have tails – the key difference between the two primates. Apes are closer to us genetically, with bonobos being our nearest wild relative, sharing 98.7 per cent of the same DNA.

However, while aware that the great apes were threatened in West Africa, I did not realise how dire the situation actually was. Bonobos, found only in the DRC, and chimpanzees are right up there on the Code Red endangered list, while the mountain and eastern lowland gorillas are even higher. They are 'critically endangered'. At the current galloping rate of extermination, most of the great ape species will be extinct within seventy-five years. The gorillas even sooner – possibly in my children's lifetime.

My initial admittedly far-fetched aim in the DRC was to establish beyond any doubt whether there were any wild northern white rhinos still alive in Garamba, but with Papa Kadima having fallen out with the government, that dream spiralled into fantasy.

It was mid-2016, and while watching all the hard work done by a lot of people at the Parc de la N'Sele unravel, I started taking a broader look at the spectacular wildlife of the DRC. As a result, I visited the bonobo sanctuary Lola ya Bonobo, which means Paradise for Bonobos, in Lingala, run by the amazing Claudine André on the outskirts of Kinshasa. This, coupled with my disturbing visit to the zoo in the city centre, rammed home the message with a sledgehammer that great apes were almost in the same freefall situation as rhinos.

Coincidentally, at the same time Paula Kahumbu, Chief Executive Officer of the Kenyan-based WildlifeDirect who had been fundamental to us during the World Youth Rhino Summit, suggested I be nominated for the Whitley Award.

The Whitley Fund for Nature is a charity that finances grassroots conservation in developing countries. Its patron is Queen Elizabeth's daughter Princess Anne, and Sir David Attenborough is a trustee. To win a Whitley Award is one of the highest accolades in conservation, and the charity has donated nearly £15 million to support hundreds of local environmental heroes in the developing world.

Paula had been a winner in 2014. The year before, the winner was a phenomenal Congolese man doing extraordinary work with gorillas.

His name was John Kahekwa.

I didn't win, but just to be a Whitley nominee was immensely worthwhile as it put me in touch with people who had won, and they are an impressive bunch. Having a wealth of first-hand experience in the DRC was why John Kahekwa's trailblazing work grabbed my attention.

As I have said, I think we were destined to meet. It certainly was also a lightbulb moment for me. The critically endangered gorilla is as iconic an emblem of the harm we are doing to fellow travellers on this planet as the rhino. The fact that only about seven hundred mountain gorilla silverbacks, or mature males, are alive today is as brutal an indictment of our criminal neglect as you will get. Future generations will possibly look upon this as being tantamount to a war crime. In my opinion, they will be correct.

It made sense that just as Rhino Art had struck a chord with a youthful, energetic and highly motivated audience, a Gorilla Art project could be equally epochal. We could replicate for the great apes what we are doing for rhinos. For a 'rhino man' such as myself, the gorilla was a similarly pivotal rallying call for the next generation.

I contacted John and we immediately hit it off. Despite English not being his first language, and French certainly not mine, we both spoke the universal language of conservation. His passion and commitment to the beleaguered creatures of his country is unsurpassed. For him, it has been a long, hard and often perilous road.

He was born in 1963 in Miti, a village in the hills above Lake Kivu, where the peaks of two extinct volcanoes dominate the skyline surrounding the huge reserve named after them – the Kahuzi Biega National Park, or KBNP as it is widely known.

The park, a UNESCO World Heritage Site, is 2316 square miles of rainforest, bamboo woodland, swamp and peat bog. It's home to a spectacular array of wildlife, including forest elephants, diverse herds of antelopes and 349 species of birds, some found nowhere else in the world. But the undoubted stars are the eastern lowland gorillas, the immensely powerful yet gentle colossi that once strode the mountain slopes in their thousands.

John was born just five miles from the park gate. It would soon become the lodestar of his life. His aunt, Agnes Bujiriri M'Rwankuba, was married to Adrien Deschryver, the co-founder and the first warden of the KBNP. Adrien, a supremely talented Belgian photographer as well as a fearless pioneering conservationist, was the first person to habituate eastern lowland gorillas to humans. John's close relationship with his charismatic uncle, who died of a heart attack in 1989 at the too-young age of forty-nine, resulted in him similarly dedicating his own life to gorillas and nature conservation.

This was not easy. His home, the eastern DRC, is unquestionably one of the world's roughest neighbourhoods. Death is part-and-parcel of daily life; rangers pitted against poachers, parks versus peoples, warlords fighting rival militias, and, particularly during the second civil war that started in 1998, basically as an offshoot of the Rwandan genocide, with Hutu killing Tutsi in an attempted ethnic cleansing where rivers literally flowed with blood. Unlike Hitler's gas chambers, neighbours hacked neighbours with machetes. You do not get closer or more personal than that.

But for John, there was no doubt that he was going to be anything other than a conservationist. He joined the KBNP in 1983 as a tracker, following two gorilla troops and identifying each animal by recording its unique nose print.

Part of his job was to familiarise these foliage-eating titans weighing up to 400 lbs to the presence of humans, as habituated gorillas would comfortably allow tourists to get close. He was so successful that he eventually was on first-name terms with at least 155 animals, the gorillas recognising him as readily as he did them. They accepted him, not quite as one of their own, but as close as any human would get.

His life changed in 1986 when a grateful tourist handed him a US$10 tip after a 'life-changing encounter' with a gorilla family. For the star-struck tourist, it was merely a low-denomination note. For John, it was an opportunity.

He got into a rusted pickup truck and drove to Bukavu, a sprawling town on the banks of Lake Kivu bordering Rwanda. There he bought ten cheap cotton T-shirts, stencilled an imposing gorilla silhouette on the fronts with the words 'I Tracked Gorillas in Zaire' (as the country was known at the time), and sold them to tourists. Profits were ploughed into printing more of the same.

In 1987 the movie *Gorillas in the Mist* hit the circuit. It was about the life of Dian Fossey, the American primatologist murdered in Rwanda, starring Sigourney Weaver. John, with his extensive tracking skills, had been hired as an advisor for the film and played a small part when the crew came to the KBNP. The movie was a massive hit, grossing close on $62 million and getting five Oscar nominations.

Suddenly John's T-shirts were in demand. After a flurry of sales, he had about US$6000 in the bank – a significant sum in the DRC.

By now John was chief ranger at the KBNP and his main headache was how to curb rampant poaching. Although the park was officially protected, that was just semantics. The reality was the constant threat from the starving people living alongside it. It was the classic Third World parks versus people situation. A growling stomach knows no rules.

John saw it from both sides. He had a foot in both camps; his much-loved gorillas facing extinction through relentless hunting and habitat destruction; his own people barely eking out a subsistence survival.

It was a brutal line in the sand. There were no winners. Poverty, lack of jobs and education forced hungry communities into the resource-rich park either to kill animals to eat, or chop down forests to cook them. How to cross that shadow line, where both animals and humans could at least coexist, was what kept John staring at the ceiling at night.

Then, in November 1993, the country's most famous gorilla was killed by poachers in the KBNP. His name was Maheshe, an impressive, hugely muscled silverback, whose face appeared on the Congolese 5000-franc banknote. Everyone in the Congo referred to the note as a Maheshe – 'this will cost [x-number] of Maheshe' was how trading was done. Maheshe was as famous and far, far more loved than the kleptocratic President Mobutu Sese Seko. His death was distressingly symbolic that the war against bushmeat poaching was lost. The irony that his face was on a banknote was more than heart-breaking. It was an omen.

For John, this was a tragedy that could not be swept under

the carpet as just 'one of those things'. A new approach was needed. But what?

It came to him in a Damascene moment after he had arrested the same poacher in the KBNP for the tenth time. As he confiscated multiple snares and a battered 12-gauge shotgun, he exasperatedly asked the man why he kept doing it.

The poacher replied it was the only way to feed his family. There were no other jobs.

John knew that. He also knew he didn't have a solution. But perhaps the poacher did?

'If you had a job, would you stay out of the park?' John asked.

The man nodded. Absolutely.

John decided to take the serial poacher at his word. There were no jobs around, so John opted for the entrepreneurial route. He would create them. The illegal hunter was taught how to carve wood, enabling him to start his own business.

That was the start of an extraordinary one-man crusade to save the continent's dwindling wildlife. It was founded on one of John's quotes that is now famous: 'Empty stomachs have no ears.'

Then, in the face of terrible human poverty, John spent the entire $6000 savings from his gorilla T-shirt business to launch the Pole Pole Foundation. Pronounced 'polay-polay', it means slowly-slowly or 'gently does it' in Swahili.

Its goal was twofold: to protect the park as well as supporting the surrounding communities. John knew the issues were inseparable.

Using the recently arrested poacher as a role model, he persuaded the other most prolific poachers in the park also to train as woodcarvers. There were forty-seven of them. For this he needed wood, so he convinced a Japanese donor to fund the planting of 10,000 saplings.

Today visitors can buy stunningly realistic figurines of eastern lowland gorillas lovingly carved by men who were once killing them.

Then he moved on to the poachers' wives, whom he knew were also either killing animals or selling their husband's bushmeat at fly-blown outdoor markets. He bought several sewing machines and taught the women to sew shirts, which they sold.

Pole Pole expanded further, building schoolrooms for environmental education and establishing tree-planting projects, eliminating the need to hack down the reserve's World Heritage forests. The young trees – mainly fast-growing eucalyptus – soon took root and were harvested by local communities for building homes, firewood and charcoal.

The success was stunning. This was working – a radical conservation blueprint, designed and rooted in Africa for the future of the continent's wildernesses.

Then it came crashing down. In 1998, the country's second civil war broke out when then-President Laurent Kabila expelled Ugandan and Rwandan forces who had been his allies in overthrowing the despot Mobuto during the 1996 civil war. The fighting claimed more than five and a half million lives and involved nine countries, hundreds of thousands of child soldiers, and satanically evil warlords commanding scores of heavily armed and drugged-up militia known collectively as *Mai-Mai*. It was, for all intents and purposes, Africa's 'First World War' and the deadliest conflict since 1945.

The brutality was staggering. All armed parties carried out a deliberate policy of genocidal rape, with the primary purpose being the total destruction of entire families and communities. Human Rights Watch described the diabolical violence towards females – some barely infants – as 'a war within a war'. Families were forced to watch as mothers, sisters, aunts, cousins and neighbours were systematically gang-raped by group after group of armed men.

The statistics are well documented and beggar belief. The *America Journal of Public Health* estimates that at least forty-eight women and girls were raped every hour with the sum total in excess of 2 million victims. But as many rural women feared social stigmatisation, the true figure is probably higher.

Under the circumstances, most would think nobody would give a hoot about wildlife conservation; but they would be wrong. In such hellish conditions, men like John Kahekwa and other loyal rangers kept a spark of humanity flickering, caring for communities in the area and the animals in the parks. At times, amid bottomless grief and horror, all they could do was refuse to surrender and drown in despair. They did their best when just their stoic presence was at times all they could offer. They stood firm against forces of demonic evil with nothing but courage and unyielding will on their side. The debt the world owes to such people is not only unimaginable, it is unpayable.

That debt has to some extent been acknowledged. Apart from winning the Whitley Award in 2013, three years later John received the prestigious Prince William Award for Conservation in Africa at the Tusk Conservation Awards. It's difficult to think of a worthier recipient.

The DRC is now rebuilding. The resilience of the amazing people of that country is simply overwhelming. I don't believe any Western society could recover like the communities who live in the shadow of the Kahuzi and Biega volcanos have.

Rwanda, where it all started, is today being tentatively heralded as the Switzerland of Africa. Rwanda is a tiny country and easier to govern. It is smaller than Belgium, once the colonial power.

Conversely, the DRC spans the majority of Central Africa, without any meaningful infrastructure in most regions. Things will happen more slowly. It has great people . . . people such as John. People who have survived everything and are rebuilding their lives.

People such as John are the future of conservation. Here's a man who used his life savings to start a charity providing tangible benefits for both desperately poor communities and the wild animals they live with, cheek by jowl. The foundation he created provides jobs, resources and classrooms for future generations.

Here's a man who risks his life in some of the wildest jungles in the world. Who has survived the most brutal war since the death of Adolf Hitler. Who never lost faith in the future of his park, his people, his animals – no matter how bleak the situation.

Here was a man I could work with.

CHAPTER TWENTY-FOUR:

At Congo's Coalface

Having made contact with John, and knowing the daunting conservation challenges he faced every day of his life, I decided that the most urgent issue was to get him some funds.

I already had a plan on how to do this. As mentioned earlier, I had previously been promised $450,000 by the International Tropical Timber Organization (ITTO) to build a media centre at Parc de la N'Sele when I was working for Papa Kadima.

After the Parc project collapsed and the ITTO told me in no uncertain terms that they were not re-routing the money to Papa's new wildlife venture, I decided instead to approach them on behalf of John. The money had already been allocated to a DRC charity so it was on the books, and there was no worthier beneficiary than John Kahekwa. I was sure the ITTO would agree, even though the initial media centre was planned for the western Congo outside Kinshasa and John was close to a thousand miles east as the crow flies. I pointed out that we could replicate in one of the most poverty-stricken areas of the DRC the amazing work that was being done with gorillas in Rwanda. Gorilla treks in that small country today are pro rata among the biggest money-spinners for wildlife anywhere in Africa.

Sadly, the ITTO turned us down. I am not sure why, but it may have been because forestry was a bigger industry in the western Congo basin and they wanted to concentrate on education there. Whatever their reasons, it was a cruel body blow for us. With that amount of money John, who was already working wonders on a shoestring, could have instead performed miracles.

It's impossible to overstate the challenging conditions that John works under. Not only is the KBNP so underfunded that rangers are not guaranteed salaries and their vehicles have no gas, John himself lives in a ghetto.

In fact, ghetto is too grandiose a word. It is instead a sprawling shanty town slithering up the hilly slopes on the outskirts of Bakavu where, for most residents, survival is a harsh existential reality. Life, in John's world, is what it is. There are no frills.

I am familiar with shanty towns. I have travelled extensively in South America where favelas without water or light fester a stone's throw from five-star hotels. We have them proliferating like weeds in South Africa where far too many people live in tin and stick shacks with little hope and even fewer dreams.

But I have never seen anything so grim as the neighbourhood John calls home. Raw sewage spills into dirt streets that are either rutted dustbowls in summer or streams of greasy mud in winter. Electricity is mainly from 'jimmied' overhead cables that feed the city centre. The crude connections consisting of wire spaghetti nests must have fried many victims, although in a shanty town such as this, that would probably go unrecorded. For most people, water has to be drawn from communal taps, sometimes miles away.

Despite that, it is obvious that John's small house has been built with immense love and pride. The construction is red-soil brick and concrete, unlike many of the other structures cobbled together with debris salvaged from rubbish tips. Inside, it is like stepping into a sphere of serenity, divorced from the chaos and grime outside. The rooms are immaculately clean and tidy, with a TV set in the lounge and pictures of awards for his extraordinary achievements hanging on the walls.

The warmth and dignity thrumming from the walls of the humble Kahekwa home is palpable. Six people live in that matchbox house; John, his wife Odette and four youngest children. The eldest two daughters live not far from me in Durban, seeking a better life in South Africa with their Congolese husbands.

It's a pill too bitter to swallow. Here we have one of the most respected men in conservation, who almost singlehandedly was responsible for saving the lowland gorillas in the DRC, living in poverty. If ever there was a snapshot of a hero foot soldier in the fight for the planet being a prophet without honour, this was it. True, he had won two of the most prestigious awards for his work, but the cash amounts were small and the awards short-term recognition. They mean nothing when he's the forgotten son in the middle of a former warzone, an area where few are interested in investing despite it being home to one of the most iconic wild creatures. He has given his life to saving gorillas and still ends up in the ghetto.

John is the Mother Teresa of conservation. He may cringe at the description, but it is true. He looks nothing like her, of course. Of average height, he's strongly built and broad-shouldered, with an easy smile and deep humour in his eyes. He's always neatly dressed and has impeccable manners. He is nature's true gentleman.

But like the saint of Kolkata, he spends little on himself. Every spare penny is ploughed into keeping the gorillas of the KBNP alive. He told me he once owned a bigger house in his home village of Miti, but sold it when the civil war erupted to keep the reserve functioning. It was an almost impossible task with ethnic cleansing spilling over the border followed by an almost decade-long war. But the profits from that house bought precious time for the gorillas caught in an apocalyptic conflict through no fault of their own.

The Pole Pole Foundation is an awe-inspiring testimony to what can be achieved with almost no resources but an iron will. Its building, just like John's home in the Bakavu ghetto, is rustic in its stark simplicity; dark with small windows and constructed with homemade bricks. Yet – again like his home – it radiates dignity and an indomitable spirit far beyond its humble appearance.

Pole Pole is the physical essence of John's philosophy. It should be a shrine, as this is where a working African blueprint for fusing communities and conservation was born – something considered self-explanatory today, but not when John implemented it. Game reserves were sacrosanct islands where human and animals were considered to be incompatible. The fact that people and animals have coexisted throughout the millennia in Africa was ignored. Separation was enforced at gunpoint. Rural people, who survived off the land just as the animals did, were excluded from game reserves, both physically and financially. Profits from wildlife went to government coffers or rich land-owners, with nothing filtering through to those who lived on the park borders and needed jobs and resources most. The resentment that fostered poisoned the concept of conservation among communities that should have been the staunchest allies.

That's changed now, thanks to people such as John. He wrote the bible for community conservation, and today most reputable reserves implement his views in some way or another, even if they have never heard of him.

For example, Amakhala has prioritised community empowerment with projects ranging from conservation awareness to AIDS clinics in our section of the Eastern Cape. We are not alone. It's happening all over the country.

The difference is that at Amakhala we are working with top scientists, veterinarians and game-management experts as well as community leaders. John is doing it mainly on his own with a handful of people such as Nikki Jones and Richard Milburn in the UK, Juichi Yamagiwa and Kawabe Tomohiro in Japan, Ariel Kedem in Israel and now myself in South Africa. We are doing our best to keep the international flame flickering.

The problem is that not only is the DRC the second poorest country in the world (just above its neighbour, the Central African Republic), it's also one of the most expensive. It is so huge that to transport goods costs a fortune, and most merchandise is imported due to the almost non-existent manufacturing sector. Many game reserves don't have enough money to run their vehicles, using them only in emergencies. The sole vehicles that have fuel or handheld radios and other bush equipment belong to NGOs.

Yet . . . the fact that Pole Pole is working despite crippling odds is proof of the strength of the project. Its core philosophy is as resilient as tempered metal. It has already survived Africa's equivalent of a world war, and is surviving almost unaided in one of the poorest and isolated areas in the world. So imagine what it could do under normal conditions?

What Pole Pole has shown is that in the most stricken conservation areas, the wrong questions are often being asked. It's not a case of 'why are people poaching?' – instead it should be 'why aren't more people poaching?'

The answer is because guys like John have shown that there is another way. A more viable way. Pole Pole is the distillation of that ground-breaking new way of thinking.

This obviously refers to subsistence poaching. The mafia-style gangs massacring rhinos and elephants for lucrative Far East markets are, of course, another story.

As well as planting thousands of fast-growing eucalyptus trees and training former poachers to be woodcarvers – whose work is being sold in Britain, Japan and Israel – Pole Pole have started a mushroom farm as the damp, humid conditions are ideal for fungi growth. Villagers grow their own produce both for consumption and market, and are no longer plundering plants that proliferate in the park.

Fish farming is thriving, providing protein-rich food from community-built ponds and negating the need to poach bushmeat. Even more ambitiously, Pole Pole is now running a spirulina farm. This nutrient-dense algae is nature's richest source of wholefood consisting of 60 to 70 per cent protein, and nearly all the fatty acids, minerals and vitamins the human body needs. Spirulina is indeed a superfood, which is vital to combat rampant malnutrition in the area.

Equally impressive is that John has managed to build an entire school where children living near the KBNP can get an education. Obviously, conservation is part of the curriculum, but literacy and numeracy are also taught from kindergarten to secondary school. Children are being prepared for productive lives, not eking out a miserable existence killing endangered animals.

The Pole Pole school dovetails with my Adidas Foundation connections, where we provide football kit and balls for youth tournaments on Gorilla Art education days. This has already proved immensely successful with Rhino Art, so there is already a pilot project in place for Pole Pole to copy.

Soon after I started working with John, we had a major 'brand' breakthrough. As he is a winner of a Tusk Conservation Award, we are now able to put prestigious Tusk logos on our Gorilla Art projects, which could have far-reaching international implications. As I keep saying, the fight for the survival of our wild places is not only in the bush with rangers and poachers shooting it out. The real battle is in the boardrooms and classrooms. In the long run, the war is going to be waged with chalk and blackboards, teaching the next generation that it is not OK to kill gorillas for bushmeat and that rhino horn does not cure cancer or impotency. At the moment, guns are vital to protect reserves but as the classroom battles start yielding victories, the gravity will shift.

Pole Pole is doing wonders in uplifting communities, which allowed us to move on to our next goal – re-establishing gorilla tourism in the KBNP. The original gorilla trek through the DRC to Rwanda was a famous bucket-list route for eco-adventurers during the 1980s and early 1990s until abruptly terminated by the war. In fact, Dian Fossey first started her work with primates in the DRC before moving to the Volcanoes National Park in Rwanda. In those days, the steady stream of visitors brought money, opportunity and, most importantly, jobs and optimism into the area. We have to get that back.

The KBNP is one of only two places on earth where the eastern lowland gorilla survives, so it should by any yardstick be a holy grail for primate trekkers. It's happening at full throttle across the border in Rwanda where it now costs $1500 for a permit to join a mountain gorilla trek – and you have to book far in advance. Rwanda's recovery from a genocidal basket case where neighbours hacked each other to death to the jewel of Central Africa is in no small measure thanks to tourism revenue generated by gorilla trekkers.

Sadly, this is not true of the DRC. In 2017, Kahuzi Biega welcomed only seventy-five visitors, despite the fact that a DRC gorilla permit costs only $400.

There is no doubt that the biggest hurdle facing John and the Pole Pole Foundation is the perception that the country is not safe.

Like most perceptions, that is now out of date. The 'African War' ended more than fifteen years ago, and if anyone wants a gorilla experience that equals Rwanda at a fraction of the price, the DRC is the place to go. I know this. I have been there. I have done it the hard way, the most basic way. I travelled on the jam-packed buses with singing locals, squeezed into taxis crossing the Rwandan–Congolese border. I have shared street food and jokes with fellow passengers, been searched by police at road blocks. I have even taken my wife there.

I have never felt threatened. Not for an instant.

But whenever I bring up gorilla trekking in the DRC with travel agents specialising in African safaris, the security issue crops up. It's frustrating, but slowly they are grasping that the war is at last over, and the trekking costs in the DRC are absolute giveaways compared to what tourists are paying in Rwanda and Uganda. Not only that, the KBNP has the biggest gorillas in the world. It should be a no-brainer for eco-adventurers and gorilla lovers.

The safari outfitters are listening, and that is all we can ask at the moment. Like everything else, it's a case of Pole Pole – slowly, slowly.

However, my trip to the DRC was now coming to an end. With Gorilla Art launched, it was time to have some fun – for Angela and me to get out into the jungle and see the magnificent King Kongs of Kahuzi Biega up close and personal.

'Come with me,' said John.

CHAPTER TWENTY-FIVE:

Walking with Giants

The big day finally arrived. I was going to meet the famed eastern lowland gorillas, the true giants of the jungle.

Not only that, I had been promised that I would get so close that I could wave at my reflection in their eyes.

To do so in the presence of John was an unimaginable privilege. Many of the primates we hoped to see owed their lives to this driven man.

As I sipped my first coffee of the morning, I was suddenly overcome by a strange sense of foreboding. Would we actually see these goliaths of the rainforests? Were they for real? Would it all be in vain – had I travelled all this way to look at vacant wilderness?

John allayed my fears. He promised we would bump into at least one of the four habituated families in the park. I was not so sure, but his unfailing optimism was infectious.

We planned to set off in the morning, although that is not always the best time. Gorillas normally feed after waking and they move fast, chewing voraciously on the vines and thick foliage that both smothers and nurtures the jungle. This means being constantly on the move, following rather than observing them.

Obviously, there is never a bad time to see gorillas, but you get more of a sense of what a close-knit, loving family they are when they are relaxing. This normally occurs late in the day, when the troop revels in the afternoon sun, basking in the forest clearings, playing with their children, and the females are de-fleaing and grooming each other. Angela and I were told there are few more inspiring experiences than sitting among the jungle creepers and watching group interaction at the highest imaginable level of affection, care and consideration.

However, an early start provided more hours in the day and as Angela and I were on a tight schedule, we didn't want to waste a second.

Before setting off, we were briefed at the Tshivanga visitor centre near the park entrance by Lambert Mongane, our KBNP ranger. Much of the information I already knew from John, but it was a fascinating background to the safari we were about to embark on.

The eastern lowland gorilla, also known as Grauer's gorilla, is only found in the DRC and is closely related to the mountain gorillas that inhabit the Virunga and Rwenzori slopes – the fabled Mountains of the Moon. However, there are small but distinct differences. The eastern lowland gorilla is bulkier with a larger, longer face and fewer nose prints. Not only that, the KBNP gorillas are the largest living primates in the world, making even the most ferocious rugby front-ranker look like an undernourished weakling.

Lambert told us that the KBNP has more than twelve gorilla families in the highland sector of the park, but only four have been habituated and can be approached by trekkers. The largest group of twenty-one animals is led by the silverback Mpungwe, but the most impressive alpha male is Chimanuka, who has a family of nineteen. Chimanuka is a gentle giant with his wives and children, rearing an unrelated orphan himself, which is almost unheard of for a male gorilla. But that's only his home life. Otherwise, he's the fearsome king of the jungle, a ferocious brawler who has yet to be beaten by any other silverback. One of the most epic fights in the park was some years ago between him and another incredible hulk called Magaruka.

Due to their massive strength and dagger-sharp teeth, gorillas seldom fight. Instead, they strut and posture, scaring off competition by displaying their enormous physical prowess, accompanied by loud hooting and the famous chest beating. This is called 'interacting' and usually stops short of violence. If a female is impressed by another male's interacting, she may leave the group to start afresh with the new King Kong Casanova.

This makes sense in the wild. There is no medical care. No hospitals. Any injury can be – and often is – fatal and the animals know that. Violence is the last resort . . . but with two immensely powerful beasts such as Chimanuka and Magaruka circling each other, dialogue failed. Violence was inevitable.

Chimanuka won the fierce fight, but fortunately it was not to the death. However, in the age-old tradition of the victor taking the spoils of war, Chimanuka seized Magaruka's wives and children, resulting in the biggest habituated family in the reserve until Mpungwe came along. Magaruka, who also lost a hand in a poacher's trap, is now a solitary male. But even so, he is a prized trekking attraction in his own right with guides taking tourists specifically to see him.

The newest family belongs to a silverback called Bonané, who has four wives and two babies. Bonané is Chimanuka's eldest son, who once had the temerity to challenge his heavyweight champ of a father. Once again, the fight fortunately was not to the death. Bonané was badly injured in the resulting punch-up and spent several days recovering, licking his gaping wounds and resting.

The plan was to hike up into the forest at the foothills of the mountains and hopefully meet one of the habituated families. We were prepared for a trek of up to four hours, taking lots of water and some food as, even with John's promises, there was never any guarantee that we would soon come across gorillas.

The highlands of the KBNP are separated from the lowlands by a road running from Bakavu to Goma, which is not ideal for a game reserve, but not unusual. We have the same situation with the corridor bisecting the Hluhluwe–iMfolozi Park in KwaZulu-Natal, which unfortunately provides a conduit for poachers. But in the KBNP, the seriously potholed Goma road is actually an advantage to trekkers as there is a network of tracks leading off it into the dense rainforest. This makes hiking a lot easier, giving Kahuzi Biega gorilla watchers a running start, so to speak.

Seeing the look of anticipation on my face, John again assured me that we would see a habituated family, claiming his wife Odette was a gorilla magnet. Whenever she is on a trek, he said, they always encountered the animals. She is his lucky charm.

Odette smiled and nodded. She mainly takes local residents on gorilla hikes, usually the wives of former poachers' or illegal woodchoppers, thanks to an innovative sponsorship scheme initiated by John. While it costs $400 to buy an international gorilla permit, John has used money from his Tusk award to subsidise the Congolese living around the KBNP. This means that for only $20 villagers who have never

217

been into the park other than to hack trees or kill creatures are now able to view up close the primates that are their nearest neighbours. For most, this is the first time they have seen these magnificent creatures alive.

The trek could have been gruelling as we were walking on an undulating mat of dense vine foliage. Every now and again one of us would fall through or slip on unseen moss-covered logs and have to scramble out of the undergrowth. Fortunately, it was the dry season, which made hiking considerably easier.

John's irrepressible optimism never left him, but even he was stunned when fifteen minutes after leaving the Goma road we stumbled across flattened vegetation, indicating gorillas were nearby. It's impossible to track the primates as the dense undergrowth covers the ground. There are no footprints in the dirt. Trampled foliage is the only indication of their presence.

'Gorilla nest,' said John, pointing to the stomped vines. 'They slept here last night. We are close.'

I could see nothing. Jungle growth, thick as a velvet curtain, swamped us and visibility was a few yards in every direction except the sky blinking through the treetops.

Then I smelt them, despite the fact that I, like everyone else, was wearing a surgical face mask as gorillas are highly susceptible to human germs. They have no immunity against flu or other nasty viruses. A sneeze can kill an entire family.

It was just a brief whiff – a wild musky odour, something I knew well in the bush. It's a healthy smell, not unpleasant in the slightest: feral, earthy and natural. Organic even. The smell of life.

I saw the canopy ahead briefly shake and heard a branch snap. Bushes were rustling. I sensed they were right in front of us, but the jungle was so impenetrable I could barely see.

One of the rangers chopped a gap in the foliage. We all craned our neck, peering through the hole as if it was a cathedral window.

Then we saw him. Sloping Neanderthal forehead, pug-face, flaring nostrils and chocolate-brown eyes with a boxer's neck cemented on tank-wide shoulders. The rest of his body was hidden by shrubbery.

I was at last face to face with the king of the jungle – yet there was no unfriendliness in his expressive features. He calmly scrutinised us as he carried on eating, stripping and shredding vine leaves in one fluid movement while delicately holding the stem as if it was a kebab stick.

He caught my gaze for an instant. He looked up, directly engaging me.

I have spent all my years around animals, some as a farmer, but mostly as a conservationist. I have travelled this continent's unparalleled wildlife and its wildernesses for decades. I know it well. I love it infinitely.

But nothing prepared me for this encounter. There is no guideline for meeting a 400-pound silverback in the wild. It's different for every individual. It's unique.

In my case, adrenaline jolted through me as if I had grabbed a plugged-in electric wire – but at the same time I felt so overwhelmed I wanted to kneel and weep. I was as wired as a junkie tripping on euphoria, yet bizarrely melancholic at the same time. The buzz, a confused blend of yin sentimentality and yang elation, was that intense.

I am not sure what triggered these vastly conflicting emotions. Perhaps it was the sheer magnificence of meeting such an iconic creature just yards away, but that's too superficial. I now think it was instead the almost holy expression in his eyes ... a look of trust and acceptance. We were on his turf, invading his space. He could have killed any of us with a single blow of his rock-sized fist. He could have tossed us onto the jungle canopy top with a flick of those slab-muscled forearms.

Yet he chose not to.

The question is why.

We humans are responsible for driving these herbivores to the brink of extinction, and yet here was this barrel-chested, six-and-a-half-foot-high hulk, fifteen times stronger than any steroid-ripped weightlifter, allowing us into his world.

Why?

It is, possibly, something humans cannot answer. I can't. The question is too painful.

Despite the fact that his anger should be cosmic in its intensity at what humans have done ... on that day in Kahuzi Biega, this supremely powerful silverback magnanimously allowed us into his world.

What have we done to deserve it?

It's enough to make strong people weep. But I should instead have gone down on my knees in the vast mat of knotted jungle and given thanks, rejoicing in the unconquerable soul of the wild, its undefinable vitality distilled to its nucleus by the immensely dignified creature before me.

'It's Bonané,' said John.

Bonané – the legendary Chimanuka's firstborn son.

The front ranger sliced through more foliage so we could view the magnificent primate better. He signalled for me and Angela to come closer. We did so, until we were only a yard away from the silverback.

Angela and I looked at each other. Her eyes were moist. We nodded at each other in perfect telepathy.

John told me such rapture affects most people seeing these stunning creatures for the first time. And John has taken a lot of people into the jungle; from billionaire Bill Gates and politician Al Gore, to humble hunters and barefoot woodcrafters who wish to pay tribute to these majestic creatures. Even the famed wildlife photographer Alan Root, sadly no longer with us, wept openly on his first gorilla encounter with John.

The rangers, most of whom come from the Batwa tribe, are so attuned to their environment that after a while Bonané barely acknowledged us. The only time he showed a modicum of aggression was when one of the rangers, while chopping off foliage so we could get clearer photos, got too close and Bonané barked at him, like a dog with the deepest bass bark imaginable. The ranger retreated half a yard. The silverback continued feeding, vigorously chewing as if his food was going to be wrested from him. It would take an exceptionally brave, or rather insanely stupid, creature to do that.

The secret of getting close to gorillas is receiving permission from the patriarch. Rangers will not allow trekkers to approach females or babies before being approved by the silverback. Homage has to be paid to the head honcho. Nobody, human or animal, can approach any member of the group without the boss's permission. If anyone does not honour the basic good manners of the jungle, they have to accept the consequences. You don't want to do that.

In front of Bonané, slightly lower down the slope, were his wives and two children. We could barely see them as they were covered in foliage, but John pointed three out; Mukono and Iragi – a mother and daughter pair who, like Bonané, had previously been with Chimanuka's group – and Siri. Females only stay with silverbacks for protection and consequently choose the biggest, most ripped specimen around for obvious reasons. Unusually, this did not happen when Chimanuka badly beat up Bonané in the gorilla equivalent of a pub brawl over women. Although the females left him for some weeks while he recovered, they later returned even though he lost the fight.

Bonané means 'the new one' as he was born on the last day of the year in 2002, and John knew him well. The familiarity between the two was reciprocal. There was no doubt that this rippling-muscle titan accepted John as a friend. Whenever John spoke, Bonané looked his way, acknowledging him.

John has adopted his own way of talking to gorillas, using a series of barks and grunts. He learned most of this 'language' from hard-earned experience, sleeping, living and eating with his wild friends. But the initial inspiration came from his pioneering uncle and mentor, Adrien Deschryver.

John told us that Adrien's method of gorilla habituation was fundamentally different to that of Dian Fossey. As he outlined in an article written for *Gorilla Journal*, his uncle Adrien always looked a gorilla in the eye and stood up straight. Dian Fossey did the exact opposite, kneeling and averting her gaze. Adrien also spoke directly to gorillas, whereas Dian communicated solely with gestures.

Both methods achieved exactly the same result. Dian's many disciples in the Volcanoes National Park across the border in Rwanda continue her legacy, while Adrien's ground-breaking work is taken to new heights in the DRC by John.

For me, it was the first time I had seen in-depth human communication with one of our closest DNA cousins, and it was a rare privilege to watch. John is a genuine gorilla 'whisperer', a word that has been horribly degraded by overuse but, like most clichés, it is true.

John said that despite all the years with gorillas in the jungle, he still loves spending hours observing their antics, the babies playing with their ever-tolerant parents, the animals grooming each other or just snoozing in the sun. To habituate a gorilla family takes years, and John has paid his dues with interest doing that. The results are amazing, not least being the obvious recognition between him and the animals. He not only knows every gorilla by name, but their dates of birth, their parents and siblings.

I watched, totally engrossed at the timeless scene unfolding before me as we followed the group ripping through the foliage gorging leaves and stems as efficiently as an industrial shredder.

Bonané then paid us the ultimate compliment. As if on cue as we were about to leave, he thumped his chest in the typical pose for which gorillas are celebrated. I have never heard flesh on bone sound so loud. A chest the size of an oil barrel pounded by a fist bigger than a brick makes a bongo drum sound mute.

I would like to think he was saying goodbye to us, but even in my most anthropomorphic moments, I know that was unlikely. He was calling his wives and children. Wild creatures are not humans. Thank God for that.

John and the rangers then offered to take us deeper into the jungle to try and find Chimanuka, the reserve's alpha male. It was an enticing proposal as Chimanuka has a fascinating history. He was the first silverback sighted after the war in the eastern DRC ended. He was like an apparition coming out of the mist as many conservationists feared all gorillas in the country had been slaughtered. Even John, who did everything humanly possible to keep the animals alive, had at one stage been forced to flee from the park at gunpoint or else his family would have perished.

Chimanuka's name means 'good fortune when you need it', which pretty much sums up the situation at the time. Before the war there were twenty-five gorilla families in the highland sector of Kahuzi Biega, and it is estimated that more than half of the primate population was massacred during the conflict. The fact that some survived was only confirmed with the sighting of Chimanuka and brought indescribable joy to John and the park rangers. Chimanuka may be the alpha male, but the symbolism runs far deeper. He is the beacon of hope for his species in Kahuzi Biega.

He's also easily recognisable as he has a tear-shaped protrusion near his outer right eye. The fact that his ID – for want of a better description – is a teardrop is poignant beyond words.

We decided not to trek further into the hills to find Chimanuka as we'd already had an overwhelmingly memorable experience with his son Bonané. Also, we didn't want to stress the animals more than necessary.

Another thing we learned from John is that the KBNP game rangers who had got us to within a yard of these impressive beasts had not been paid for eight months. Appalling as that is for people living below the breadline, eight months without salary for them is not unusual. The record is three years. The commitment of people who keep working with no guarantee of regular salary, just the unyielding love for the creatures in their care, is humbling. They are the true wildlife champions of the planet.

After the trek, John pulled me aside and said he believed this was God's way of bringing us together. I could not dispute that. The Democratic Republic of the Congo, the clichéd Heart of Darkness, is against all odds one of the most exhilarating lands in the world. And with Gorilla Art now established, coupled with the astonishing work John was doing with Pole Pole, I was confident Kahuzi Biega would re-emerge from its black hole as a holy grail on the gorilla-trekking route.

I would be returning often. I owed that to Bonané's immaculate hospitality and generosity of spirit.

He had touched me to the core.

But now I needed to get back home to South Africa where the land expropriation issue was hotting up and ready to explode.

CHAPTER TWENTY-SIX:

Land Made in Anger

It was as gripping as a Wilbur Smith plot.

Rampaging elephants, big-game hunters, poachers, hard men, wild country and angry tribes, set on a sprawling backdrop of the vast African bush.

Yet this was no armchair adventure fantasy by the continent's bestselling action-adventure writer. It was happening on a former hunting concession outside a tough coal-mining and cattle-ranching town called Vryheid in KwaZulu-Natal, about 150 miles from where I lived.

The background story revolved around a long-standing feud between an Afrikaner hunter who had owned the large concession, or game farm as South Africans call it, and the disgruntled people surrounding his land.

In some ways it was a parable of old and new South Africa. The hunter – like apartheid – had recently died, but the anger was very much alive.

His land was called Mawana. It was a name that was to dominate the next few years of my life.

One of the core disputes in modern South Africa is land expropriation. The issue is not a theoretical one of whether land should be expropriated from white South Africans, it's already happening. It's also as volatile as gelignite, but sane heads have kept the tinder from flaring. So far, that is.

The simple reality is that the minority white population still owns most of the land, and this is a festering scab for many. How to resolve it amicably, and without imploding the economy, is one of the country's existential dilemmas.

In many instances, agricultural land was seized by the apartheid regime in the middle of last century to make way for large-scale commercial farming. This caused incalculable anguish as communities were uprooted and relocated elsewhere.

But not in most cases, and that is what makes the situation so complicated. It's not a cut-and-dried issue of returning stolen lands to historical owners. Thousands of white farmers have legal title deeds to lands that were never populated beforehand, and to merely evict them out of hand could ravage the economy. It's not me saying that – a cursory glance across the border to bankrupt Zimbabwe tells the story far more eloquently than I could.

As far as possible, land claims have been run on a willing buyer and willing seller basis. 'Willing' in some cases may be too strong a word, but most white South Africans have accepted it as reality. Farmers were informed that surrounding communities had a right to their land, and the government would buy the farmers out. Due to disagreements on the value of the property, this was often a slow process and led to much frustration among both farmers and the black communities. I experienced this first-hand with my banana farm, which took eight years to conclude.

In 2018, the government under new leader President Cyril Ramaphosa took this a step further, decreeing that in future white land could in various circumstances be legally seized without compensation. President Ramaphosa was egged on in this regard by Julius Malema, the firebrand leader of the Economic Freedom Fighters Party, whose standpoint is that all land belongs to black South Africans. In essence, he believes whites have invalid title deeds to stolen property – this is a convenient dog-whistle to blow at tub-thumping political rallies. Mr Malema is consequently demanding that white Africans be kicked off their properties or face dire consequences. To say this has caused concern, both internally and internationally, is understating the case.

I steer clear of politics as best I can. But land expropriation is something that nobody with an interest in the future of Africa's wildlife can ignore. As with the Tugela reserve, where we helped with anti-poaching horse patrols, a lot of wild land already stocked with game has come into community owner-ship. The new owners – mainly rural villagers – are not able at even the most basic level to deal with wildlife and land management issues, and the results have tragically spoken for themselves: rampant poaching, fallow lands, massive soil erosion and environmental degradation. Even worse, destitute communities promised jobs and prosperity have instead sunk even deeper into crippling poverty.

Through no fault of their own, rural people eking out a subsistence living are obviously not equipped to run sophisticated commercial ventures. They also don't have the resources and the know-how, or access to bank loans. Thus to hand over a functioning tract of previously white-owned wilderness stocked with a healthy population of animals to a community with no experience in running it is not going to work. Even worse, when communities were promised assistance, they were often victims of confidence tricksters claiming to be wildlife 'experts' and surreptitiously running hunting safaris or selling trophy animals to unethical gunmen for their own profit.

The consequences are depressingly apparent. In some appropriated wild areas, all rhino – both black and white – have been shot out and their horns shipped to the Far East, while hunting antelope such as nyala, kudu and impala with packs of dogs is rampant. Snares, slowly garrotting animals to death with extreme agony, are as plentiful as ticks.

Economically, socially and environmentally, we have seen one disaster after another. Vast tracts of hugely valuable land that could benefit entire communities were instead reverting to rack and ruin.

Agriculture had been touted as one option, but most rural people are subsistence rather than large-scale farmers. To turn wild bushveld into farmland takes heavy machinery and significant capital – both conspicuously absent in poor communities.

Yet there is a solution. A glaringly obvious one: wildlife tourism. One does not have to be an economic genius to grasp that this is the most viable industry, as the product – wild animals – is already in place. Establishing game reserves owned by local communities and run by ethical wildlife managers with the help of reputable conservation organisations is a no-brainer. The lion's share of the profits will be ploughed into the community, establishing schools, clinics, halls, playgrounds and transport services, and also providing jobs in the lodges as hospitality staff or in the field as game guards and rangers. Spinoff industries would include shops and other cottage industries selling curios and woodcarvings to tourists.

In other words, the reserve would be run along similar lines to John Kahekwa's community-conservation Pole Pole project. As John had shown in the DRC, in many cases it was not just a viable solution, it was the only solution.

This is where conservation NGOs come in. We are specifically geared for this type of development. We have the wildlife management skills and people – dare I say it? – such as me who have grown up with rural communities. People who know their culture, customs, beliefs, superstitions, myths, legends and tribal histories. And above all, we speak their native tongues fluently. If ever I needed to get down on my knees and say thank you for being brought up in the bush and the mud huts of my *kwedini* friends, learning folklore from tribal elders and inspirational teachers such as Alistair Weakley, now was the time. This has immeasurably equipped me for a spiritually rich life on this continent that I would never have received anywhere else. I have fostered trust and friendships that I could not have made without my rural African background. I may not have fancy university degrees in conflict resolution or the legalities of land acquisition, but my entire life – from marketing goats in the outbacks to fostering conservation in the remotest schools – has been forged in Africa's fire. My upbringing has been a preparation for a vivid life in Africa that no classroom or lecture hall can teach. I am supremely blessed.

This is what I believe I bring to Project Rhino. We are always on the lookout for range expansion, finding wild land to stock with wild animals as their habitats shrink with galloping urbanisation. One of the key participants of Project Rhino, the Black Rhino Range Expansion Project (BRREP) has been tasked by the World Wildlife Fund for Nature to find bushveld suitable for up to twenty black rhinos every year. To make it even more challenging, the WWF has stipulated that it will only consider properties that are at least 70 per cent owned by communities; the habitat has to be perfect; and the land has to be a minimum of 20,000 hectares and safe for rhinos.

That is no small ask. To find a piece of wilderness that fits all three criteria is the perfect storm in range expansion. We may have one with the right percentage of community ownership, but not enough land. Or we may find enough land, but it is not ideal rhino habitat. Or it may not be safe – a big deal with the rhino wars raging.

As good luck would have it, I came across a perfect spot in an area about 30 miles outside Vryheid. A chunk of pristine bushveld called Mawana Game Reserve, which already boasted impressive wildlife, although all rhino in the reserve had been shot out. But it was superb wild country, and that's what piqued my interest. This was the future; we would restock this sublime bush with rhino again and guard them around the clock. In fact, BRREP's internationally renowned veterinarian, Dr Jacques Flamand, said it was among the 'most perfect rhino habitat' he had seen for many years.

But as bad luck would have it, Mawana was fraught with problems. And the deeper we dug, the more we became bogged down in the classic Wilbur Smith plot that I outlined earlier, with life imitating art instead of the other way around.

The owner Kerneels van der Walt, a tough old Boer who solved problems, perceived or otherwise, with his ham-sized fists or rifle, had died in 2017, but his ghost still haunted the area. His mindset had been rooted in apartheid South Africa and he had sown such animosity and distrust with the local communities that reaching any solution needed the wisdom of a sage and the patience of a saint.

To further complicate matters, mixed into the strange brew of antipathy was a large herd of elephant that could be killed. The animals had escaped from Mawana Game Reserve and were now either on tribal lands or nearby. I heard that a destruction application had been lodged and seven bulls already shot – at least one in less than transparent circumstances.

Not only that, to my astonishment the Mawana elephants potentially in the sniper's sights were directly related to the herd we had stocked on Amakhala fifteen years ago.

This was getting personal.

CHAPTER TWENTY-SEVEN:

Wild at Heart

I first heard about the escaped Mawana elephants facing a potential death sentence from my friend Andrew van Ginkel, an environmental online investigation specialist.

He had been instrumental in exposing Phillip Hathaway during the Atherstone reserve saga, sifting through the at-best misguided appeals to stop the 'culling' of eighty elephants.

At face value, the Mawana elephant situation was worse than what we had been up against at Atherstone as a written application to shoot fifty-three animals had already been lodged with Ezemvelo KZN Wildlife. Andrew stressed that I had to move fast.

I phoned Johannes van der Walt, the eldest son of the late Kerneels and chairman of the Mawana Game Reserve Trust, the land his father bequeathed to his heirs. He confirmed the existence of the destruction permit and said the application had been lodged by one of the reserve's neighbours who'd had property damaged. There was also concern about the threat to human life.

I did not know about Kerneels's fierce reputation at the time, and I would never have guessed it by speaking to his son.

SAVING THE LAST RHINOS

Johannes was polite, courteous and an altogether nice guy. He also was an executor of Kerneels's estate and extremely concerned about the rapidly deteriorating situation, saying he had little idea of what to do with the family's elephant herd that was now on the run.

Providing a brief background, he said since his father died the reserve had been neglected as he and his one sister lived in Pretoria, while the other brother and sister lived in Vryheid. Consequently, there were no family members on the reserve to manage the property. As a result, the electrified boundary fences had not been maintained and in some sections poachers and cattle thieves had cut through the wire. The elephants had consequently bulldozed through the breaches. They were thought to be somewhere in an almost inaccessible valley east of the reserve.

The reserve was now basically a free-for-all. With the fences derelict, illegal hunting from some of the surrounding communities had intensified so radically that wildebeest were slaughtered solely for their tails. These would be sold as *muti*, or medicine, to *sangoma* traditional healers for 2000 rand. In other words, this was no longer subsistence poaching by hungry people. It was a business.

Even worse, it was not just impoverished peasants wreaking havoc, Johannes said. White hunters were also allegedly bringing in top-dollar Russian and American clients on questionable big-game permits, or paying bribes to get their customers in for the kill. As I said, this was straight out of an adventure novel, except it was frighteningly real.

Wildlife authorities investigated the neighbour's complaint and confirmed the elephant damage. As a result, the Mawana Game Reserve owners had been served a non-compliance notice by Ezemvelo KZN Wildlife instructing them to repair their fences and bring the elephants back into the reserve. Failing that, they had to find suitable alternative accommodation at other reserves. Non-compliance of this carried a maximum jail sentence of ten years.

Johannes was at his wits' end. He said it was almost impossible to carry out Ezemvelo's demands as the reserve did not have the funds to repair the expensive big-game fences. Nor did they have the money to relocate the elephants, which involved hiring helicopters, highly skilled bush pilots, heavy-duty trucks and teams of game experts and veterinarians.

Mawana's main source of income had also dried up as all hunting was now prohibited due to the non-compliance notice. The only current funding came from renting out an area of the farm to the Inkawu Vervet Project, a programme studying monkey behaviour run by the Université de Neuchâtel in Switzerland.

It was classic catch-22. They had to spend money to fix the problem, but the problem prevented them from getting money.

Johannes said he had phoned every contact in conservation he knew, including the International Fund for Animal Welfare (IFAW). It was to no avail. All had professed shock at the destruction application, but none had been prepared to come to the reserve and assess the situation first-hand.

So I did.

I arrived at Mawana two weeks later, little knowing that this would evolve into one of the most absorbing conservation projects in the country, combining bitter land claims, fascinating history, big-game hunting concessions converted to pristine conservancies, and unbelievable potential for large-scale range expansion for my favourite animal – rhinos.

At that initial meeting I met the Mawana owners and some of their neighbours. I also made contact with the communities, who treated me with hostile suspicion when they heard I had come to discuss the Mawana situation. It was then that I discovered the antagonism between them and old man Kerneels prior to his death. They claimed he had been responsible for chasing their people from his property, confiscating their cattle and generally treating them as unwelcome invaders rather than neighbours. As a result, they had little interest in stopping the current wave of poaching or transporting of stolen cattle over Van der Walt lands, and even less in reaching a deal with Kerneels's heirs. Their interest appeared to be in claiming all the land as theirs, and it seemed they might succeed.

Kerneels had been a giant, standing substantially north of six feet tall and built like the proverbial brick outhouse. He was a hard man of the bush with an explosive temper, and not averse to using physical force. This was not necessarily an anomaly in the more remote parts of South Africa. In fact, Kerneels was very much a product of his times, a throwback to the 1950s and 1960s, and an avid adherent of rigid apartheid dogma, regarding his neighbours, both black and white, with disdain.

In return, the Zulu communities hated and feared him. I was told that over the years he expelled at least two hundred people from his lands and claimed he had specifically brought in hyenas to kill their cattle. Hyenas are indigenous to the area. Kerneels never 'imported' them. But by claiming that he had further inflamed the situation.

I have no idea how true all of this was, but am repeating it as this turbulent history would be a major thorn in any negotiations. Whether it was fact or fiction did not matter. The reality that it was a running sore was something we could not ignore.

However, one thing not in dispute was that Kerneels wasn't what you would call 'a people person'.

But community grievances could be dealt with later, and a big plus in our favour was that, unlike their father, the Van der Walt siblings were keen to forge better relations with their neighbours. That was concrete stuff with which we could work.

I temporarily put the age-old feuds aside and instead focused on the more pressing current problem: the elusive runaway elephants. We had to find a solution to that fast or else the destruction permit could be validated.

My main contact with the Van der Walt family at Mawana was Beyers Coetzee, a remarkable man married to Kerneels's eldest daughter, Una. We were walking in the bush, chatting about the situation, when out of the blue he mentioned that the original elephants of the fugitive herd had come from the Phinda Private Game Reserve in 2003.

I stopped dead in my tracks. From Phinda? In 2003?

Beyers nodded. It was then that the penny dropped. By supreme coincidence, these animals were part of the same herd we had darted at Phinda almost fifteen years ago to move them to Amakhala. I had been closely involved with that, organising a German TV crew to film the action, which was viewed by almost seven million people. Ten of the twenty-two elephants were trucked to Amakhala to become part of our core breeding stock, and the rest obviously went to Mawana.

How coincidental was that?

To bring it even closer to home, Beyers told me one of the Mawana matriarchs had no tusks, a rare phenomenon originating from the Gorongosa reserve in Mozambique. The original Phinda herd were survivors of a mass slaughter during the RENAMO–FRELIMO civil war that turned Gorongosa into a massive abattoir to feed the guerrilla troops, as well as a macabre 'supermarket' for blood horn and ivory exchanged for weapons.

As a result, directly or otherwise, an increasing number of tuskless elephants were born. It was as if nature knew that an elephant's ivory was the biggest threat to its life, and so fast-forwarded a few evolutionary eons to spawn animals without tusks. This is speculation, of course, but certainly makes sense.

We also had a tuskless elephant at Amakhala, a charismatic female we called Ogre, which is a gross misnomer for such a gentle creature. When I heard about Mawana's tuskless cow, I knew there was little doubt that the two herds were genetically related.

It was now personal.

Beyers and I started drawing up a plan. First, we needed to determine exactly how many animals were involved. Without that, we would have problems sorting out logistics, particularly regarding relocation.

To my astonishment nobody, including the Van der Walts, seemed to know.

Beyers said his father-in-law loved elephants and four more bulls were added to the original group of twelve in 2005, which boosted breeding numbers. However, since then at least six elephants had been shot by big-game hunters, paying up to a quarter of a million rand in killing fees. The seventh, a giant bull, had been shot by a neighbour, apparently in self-defence.

After breaking out, the herd had split into two groups. Beyers said the most likely estimation of the combined total was about twenty-seven elephants, including full-grown bulls and *askasris*, young males, that are no longer an integral part of a herd. Elephant herds consist entirely of adult females and children. Once a male reaches puberty, he is evicted.

The bottom line was that this figure was little more than an educated guess. The countryside was too impenetrable and it was almost impossible to do a completely accurate game count. But by calculating births with the lengthy gestation period of an elephant (almost two years) as well as deaths at the hands of hunters, the overall number was unlikely to be more than thirty.

Which begged the question: why were there fifty-three elephants earmarked for death on the destruction permit? In other words, at least twenty-three animals that apparently didn't even exist?

Something was radically wrong here.

Beyers shrugged, but I guess he was thinking the same as me. Mawana was – or had been before the non-compliance notice – a prime hunting concession. Big-game hunters knew the area well. So imagine if the word got out that fifty-three animals were legally earmarked for destruction, when the true total of the herd was considerably fewer? Although the shooting of the animals would be closely monitored by Ezemvelo KZN Wildlife, the fact that the figure appeared to be hugely exaggerated showed there was a possibility that animals not part of the Mawana herd could be hunted fraudulently under the guise of a legal permit.

This was potentially worth millions. But perhaps I was being cynical.

Whatever the truth, there was little doubt that in the wild lands stretching to the horizon before us, elephant hunters could get the once-in-a-lifetime trophies they craved under a flimsy veneer of legality. Sadly, there are some who would be prepared to pay big money to do exactly that.

Now that we suspected that the destruction application was grossly inflated if not fraudulent, we urgently needed to get a more precise count.

I knew just the man to do it: elephant expert Kester Vickery from Conservation Solutions. Kester arguably knows more about elephant behaviour than anyone else, and elusive behemoths in impenetrable wildernesses are his speciality.

Two weeks later, Kester and I arrived at the reserve in a chopper flown by an old friend of the Fowlds family, Jason Fischer. He comes from the Eastern Cape and is a brilliant pilot. He has to be, considering that bush pilots are the modern equivalent of the old-school barnstormers, doing one of the most dangerous jobs in the world.

For the next three days we criss-crossed above the treetops over some of the most untamed land I had ever seen. It was truly magnificent. There, in the rampant thickets, groves, wood-lands and head-high savannah grass, we somehow managed to count three elephant groups – not two as Beyers had thought – totalling twenty-four animals. This tally was three below Beyers' estimate of twenty-seven, and it was possible we missed a couple. But there was no way that the three herds extended to fifty-three animals.

I could see Kester and Jason were as stunned by the glorious vista of feral Africa below us as I was. We had stumbled onto a conservation jewel, offering the perfect range expansion possibilities that conservationists like us dream about. Mama Africa was once more offering a prime opportunity; a glorious slice of pure bushveld as it once was. I was thrilled to my bones.

That was the good news. The bad was the reality that this huge tract of feral real estate also presented us with an overwhelming problem – how to get the elephants home. It would be logistically difficult, if not impossible. Even if we could track down all the elephants and dart them from the sky, we could not get the multi-wheeled, heavy-duty trucks and trailers into that dense bush to load them up and move them. The bush was also too solid to herd the animals accurately towards Mawana by flying choppers directly at them, hovering just above the treetops and spooking them towards the reserve.

Instead, Kester said the best option was to put electronic collars on the matriarchs to monitor them. At least that way we would not have to fly for three full days to find them again. Also, if there was further property damage or threat to human life, we would be able to react quickly.

An electronic collar has an embedded GPS tracking system and is bolted on with lock-nuts and clamps as the giant animals throw them off so easily. They are not cheap. This was financially beyond the means of the Mawana management and myself, so I phoned Dereck Milburn of Elephants, Rhinos & People. ERP's core mission is to preserve and protect southern Africa's wild elephants and rhinos through job creation in nearby rural communities. This, I thought, would be right up ERP's street.

It was. But Dereck stressed that we had to get the community fully on board. This was bigger than all of us. Somehow we had to find a way through the morass of historical bad blood with Mawana's neighbours to reach an amicable agreement.

While waiting for the collars to arrive, Beyers and I got talking about the destruction permit. Who was this Mr Sithole, the man whose name was on the application? And where did he get the figure of fifty-three from? It seemed so definitive.

Not fifty or sixty, but an exact amount. Fifty-three. And we knew it was wrong.

We then hit the first of many anomalies in the Mawana elephant saga. It turned out that the applicant was Dlemeveni Mfana Sithole, the sole large-scale black commercial farmer in the area, a highly respected, self-made millionaire. I had met him before, an immensely humble yet inspiring guy. I thought there was no way he would put a fraudulent figure on an animal destruction permit, and so confronted him about it.

It turned out he hadn't. Not only that, he flatly denied that he'd even applied for a permit. While he had suffered damage from marauding elephants and reported it to Ezemvelo KZN Wildlife – as Johannes van der Walt told me – he never demanded that the animals be destroyed.

It was true that Mr Sithole was a keen hunter, but he was also an ardent conservationist and vehemently anti-poaching. He grew up as the son of a labourer on the farm he now owned, buying out the former white landowner. All around his property were signs saying dogs will be shot on sight as, in his area, most of the poachers were tribesmen with packs of scrawny lurcher-type hounds.

The destruction document was again checked. The identity number matched Mr Sithole's. But the signature didn't. Mr. Sithole's ID document had obviously been 'borrowed'.

The application with the fake signature allegedly originated from a gun shop in a town in the KwaZulu-Natal Midlands, a hundred or so miles away. Further investigations revealed that there also appeared to be a white farmer and an allegedly compromised District Conservation Officer involved in submitting it.

As a result, Ezemvelo KZN Wildlife flatly refused the destruction application and continued its investigations into who was behind it. We were just happy that the first hurdle had been cleared. The elephants were not going to be destroyed.

But at the same time, the problem was not going to disappear. If anything, considering the dense bush the animals were in, it had intensified. Fulfilling the two options stipulated in the non-compliance notice – repairing the fences and moving the elephants back to Mawana or finding them another home – appeared even bleaker.

There was another option, one that came to me while flying with Kester and Jason over the rugged hills and valleys where the elephants were lurking.

'What about creating a bigger park?' I asked Beyers when I next saw him. 'A greater Mawana park that would include the area the elephants are already in? Then we wouldn't have to move them.'

Beyers smiled. 'Funny you should say that.'

CHAPTER TWENTY-EIGHT:

Loziba – An African Dream

Kerneels van der Walt originally bought the 6000-hectare farm called Mawana in 1989, extending the area to about 10,000 hectares by buying an adjacent piece of land a few years later.

He started farming cattle, but soon realised that the country-side was too harsh, while stock theft from bandits in surrounding areas made the Wild West look like a bedtime story. He then started game farming, running hunting safaris as the land teemed with wildlife running free in the untamed bush.

Herds of blue wildebeest, nyala, kudu, red hartebeest, impala, zebra, waterbuck and giraffe roamed the area. At one stage there were also white rhinos, but the last remaining animal was slaughtered and dehorned by poachers in 2014. The only testament that rhinos had once thrived on these lands is a faded photograph of a defiant cutlass-horned beauty hanging on the wall of the Mawana lodge's bar. That was something I hoped to change. I vowed that there would be living rhinos gracing this ancient landscape once more.

However, despite the obvious enormous potential for conservation, it was tough being an optimist. Much of this wilderness had already been redistributed to the tribal communities, while the rest was mostly lying vacant, awaiting claim court rulings. The new owners were not sure what to do with their newly acquired properties. This was understandable. They had received title deeds with no guidance, advice or skills training.

On unresolved land claims, some farmers had simply packed their bags and left, not willing to continuing planting, ranching or ploughing capital into crops or cattle that they would soon lose.

With the famers gone, employment was in freefall. Anyone with a modicum of secondary-school education left for the towns, mainly to Zululand's commercial hubs of Empangeni and Richards Bay, or even as far afield as Johannesburg.

In short, it was a mess of epic proportions. Most of the agricultural land lay fallow while the bushveld was a poacher's paradise.

But it didn't have to be like that. The glittering prize was the wildness of the land itself. There are pitifully few areas left in South Africa where someone is able to sit around a glowing log campfire at night and not see a single electric light blinking on the horizon. The greater Mawana park that we were proposing to create was exactly that. It had that matchless primordial solitude, an asset impossible to quantify. Finding prime wild habitat suitable to convert into giant conservancies is rarer than gold dust, and at the rate wildernesses are being gobbled up by urbanisation, it's far more valuable.

The sun was setting, warm on my skin, as I creased my eyes to cut the glare and gazed over the savagely beautiful valley where the elephants were hiding. I knew in my marrow that Mama Africa was once again providing precious opportunities.

Despite the current morass with land distribution, this was a patch of prime real estate that would bring opportunities for all who lived here. And the animals would not suffer.

It also ticked other boxes. The proposed biosphere area was almost 40,000 hectares of wilderness, and it was more than 70 per cent community-owned.

About 2000 hectares of the original Mawana reserve were not earmarked for appropriation, so would remain the property of the Van der Walt heirs. Although this was a fraction of the original land mass, it was an integral jigsaw piece of the overall biosphere. Of far greater significance was that the Van der Walts owned all the animals in Mawana, including those in areas being appropriated. Land claim courts only take into account fixed assets, which do not include game.

This meant that for the greater biosphere to be fully stocked, we needed the Van der Walts' wild animals. Otherwise they would be legally entitled to move them elsewhere and we would be left with barren bushveld.

Consequently, the Van der Walts were in a key negotiating position. It also gave them a vital stake in the area. In any future development, they would be founder members and shareholders. For the greater conservancy to work effectively, the Van der Walts needed the tribal communities as much as the communities needed them.

Cooperation was key. Thankfully, Johannes and Una, the Mawana Game Reserve trustees and executors of Kerneels's estate, were almost the exact opposite of their father; friendly and a pleasure to work with. Johannes is a brilliant naturalist and has several PhDs to his name, while Una is solid and reliable, wanting to bring prosperity into the area where she grew up. Equally vital to the setup was Una's husband Beyers, whom I sensed straightaway would be our biggest ally in reaching a solution. He is an impressive guy all round, immensely knowledgeable about the bush, a world-class bow hunter and tracker as well as committed conservationist. Above all, he was determined to establish a legacy from his wife's family lands that would finally pull down the curtain on the area's antagonistic history. Without him, I was not sure whether we could pull off this massive gamble.

However, we could not ignore the fact that the Van der Walts were steeped in a serious legal quandary. They faced jail unless they did something about the roaming elephants. And soon. If one of the elephants killed a human, this could spiral out of control.

A clearer picture was at last starting to emerge. The creation of a greater park would in effect be a threefold rescue plan. It would rescue the surrounding communities from their current stranglehold of no jobs and dismal poverty; the Van der Walts from a jail sentence; and the elephants from possibly being shot. None of the involved parties could afford failure.

Bizarrely, whichever way you looked at it, the stakeholders had in fact been brought together by the elephants. If the herds had not broken out of Mawana, it was unlikely that the community would be speaking to the Van der Walts, and vice versa. It would all be in the hands of multiple lawyers in land claim courts, which could take years – perhaps even decades – to finalise. The stakes were high, but the elephants had, unwittingly of course, brought all parties to the negotiating table.

Then to my surprise I discovered that my idea of creating a massive game park was not new after all. I was told it had actually been proposed more than twenty years ago.

'By whom?' I asked.

Beyers replied, 'Kerneels.'

I laughed out loud. Kerneels, the most feared man in the area? Who had evicted tribesmen from the land, confiscated their cattle and antagonised virtually every neighbour? How could he be the visionary of something that essentially involved widespread community cooperation and trust?

'Look at this,' said Beyers, pulling out a tattered folder stuffed with piles of paperwork.

It was the original greater park plan, drawn up by Kerneels himself. He had compiled a blueprint way back in the 1990s. He even had a name for it: the Thaka Valley Game Reserve.

I studied it, excitement growing. It was not exactly as we envisaged, but close enough for us to copy and improve where necessary. As a result, we had an instant basis for a working business plan. There were also detailed maps giving us a head start as much of the topographical research had already been done.

The greater reserve boundaries included the deep valley where the elephants were holed up. As a result, this would not only create a park with three of the Big Five – elephant, lion (two escapees from the nearby Hluhluwe–iMfolozi Park) and leopard were already there – it would also solve the elephant problem to everyone's satisfaction.

A Big Five reserve had been Kerneels's vision before he died. The only problem with the old man's dream was that there were no takers – white or black – who wanted to share it with him. As a result, it withered on the vine.

Now was the time to revisit it.

The next step was getting the community to buy into the idea. This would involve numerous meetings with a myriad *amakhosi* chiefs and *izidnuna* headmen as well as the communities themselves, but that was something we were geared up for. Community–conservation liaison was my speciality. That would be my main role. However, I knew this would take my negotiation skills to the wire thanks to the overt suspicion I had already encountered. First, I had to get the message through to the chiefs and headmen that the Van der Walts were extending an olive branch in a manner that possibly could never be repeated.

I couldn't help smiling. Truth is always stranger than fiction, perhaps more so in Africa than elsewhere. By some bizarre quirk, the man rightly or wrongly held responsible for much of the conflict in the area was also the man who years ago had come up with a solution.

Then we discovered something that upped the ante even more. This greater park was once home to one of Africa's greatest warriors, Mzilikazi, the founder of the Ndebele tribe and arguably second only to Shaka Zulu as a military genius.

Mzilikazi's kraal, or homestead, was included in the greater park. You can still see the remnants. We were standing on historical dynamite. It made conserving the area into something for the future even more urgent.

This was also where Mzilikazi grew up and his clan, Khumalo, is today one of the most predominant tribes in the area. It is also possibly where he met Loziba, the woman who became his favourite wife and queen.

We now had a name for the new park. No longer would the word 'Mawana', and all the historical baggage that carried, be used.

It would be called Loziba.

CHAPTER TWENTY-NINE:

The Warrior Ethos

It is not for nothing that the most famous book on Mzilikazi is called *Path of Blood*.

In those three words, the author-historian Peter Becker encapsulated the extraordinary reign of one of Africa's most charismatic leaders.

Mzilikazi was born around 1790, the son of an influential chief in the Mkuze area, but spent much of his childhood on what became the Mawana game farm. It was where he spent his formative years, and remnants of his ancient rock kraal are still standing on what local Zulus call Mzilikazi Hill.

Another link is that one of the mountains in the area is called Mizilikaat, next to the higher Mawana Mountain, after which the Van der Walt game farm was named. However, *mawana* is not a Zulu word. Instead, it comes from Zimbabwe and is the Ndebele name for baobab tree, literally meaning an 'alone standing big thing' – an apt description of Mzilikazi, even though his name translates into English as 'The Great Road'. Which is equally apt.

The Ndebele tribe was founded by Mzilikazi close on two centuries ago in the South African Highveld and they later moved north of the Limpopo River, about 900 miles from his birthplace. Is it too much of a coincidence that the biggest hill in the area has a Ndebele name? That is open to debate, but the circumstantial evidence is strong.

However, there is no debate that Mzilikazi is a giant in southern African history.

As the leader of the Northern Khumalo clan in Zululand, he initially swore allegiance to Shaka and soon became a key advisor to Africa's fiercest warrior king.

In fact, Mzilikazi's fearlessness in battle resulted in him becoming Shaka's favourite general. However, that was not how the Khumalo clan leader saw himself. He wanted his own kingdom.

His chance came when Shaka sent Mzilikazi's regiments to attack the Sotho chief Ranisi. After routing the Sothos, Mzilikazi's men rounded up the vanquished tribe's cattle. But instead of delivering the spoils of battle to Shaka, Mzilikazi kept them for his own clan. The punishment for such disobedience was a particularly painful death, something Shaka aimed to inflict on his once-favourite protégé. Consequently, Mzilikazi and his followers fled northwards into the Highveld hinterland, which later became known as the Transvaal, and today as the Gauteng and Limpopo provinces.

This is where the path of blood started. And it flowed wherever Mzilikazi went, pillaging all before him and giving young captured men two options: fight for him or die by evisceration. Women prisoners became concubines, breeding the next generation of hard fighting men.

But despite being a ferociously cruel general, Mzilikazi was also one of the continent's first statesmen. Instead of having an army of demoralised and disgruntled 'slaves' from other tribes, as one would have expected, he instead fostered a loyalty unparalleled at a time when tribal warfare was borderline genocide. Soon Mzilikazi had the largest and most lethal standing army in the southern Africa interior, feared by all. The mere whiff of one of his regiments advancing resulted in entire villages stampeding in terror.

A new nation was born. The Sotho-speaking tribes of the Highveld called Zulu (or Nguni) speakers from the east coast 'maTebele', which became the generic name for Mzilikazi's followers, even though they referred to themselves as Ndebele. His policy of 'persuading' captives to join his regiments resulted in as many ethnic Tswana and Sotho warriors as Zulu, but they were now all Matabele, speaking the Ndebele language. And Mzilikazi was their all-powerful king; their lodestar. This was possibly South Africa's first rainbow nation, albeit in language and culture rather than colour.

For the next decade, Mzilikazi was elevated to almost godlike status, displacing all other tribes and anyone who dared to cross him. Most historians call this period the *Mfecane* – the crushing. Another interpretation of the word is 'scattering', resulting in devastation, community displacement and mass murder on a colossal scale. The *Mfecane* story is controversial, as some African historians consider it to be a Eurocentric interpretation and subliminal excuse for whites moving onto land they claimed was uninhabited. But still, estimates of the number of people killed during the *Mfecane* average out at 1.5 million, and there is no dispute that it actually happened. Global, let alone African, populations in the 1830s are a fraction of what they are today, so the impact of the death of possibly up to 2 million people is almost unimaginable. I am no historian but, significantly to me, that estimate is roughly similar to the number of people killed in the Rwandan genocide that so brutally impacted on my friend John Kahekwa in the DRC.

There is also no dispute that the area became so sparsely populated that when the Voortrekkers, the Boers leaving the rule of the hated British in the Cape, arrived in 1836 they initially encountered little or no opposition.

Ironically, Mzilikazi had helped his eventual downfall. All tribes that could have hindered the Voortrekker migration northwards had been dispersed by the king himself.

The Voortrekkers and Matabele clashed repeatedly, with the numerically tiny but heavily armed, highly mobile and deadly accurate Boer riflemen eventually achieving what no other tribe had – forcing Mzilikazi's warriors armed with spears and cowhide shields to retreat.

There, at his headquarters at eGabeni along the Marico River, a nine-day battle erupted, with the Boers destroying the settlement, as well as a string of Matabele camps on the river banks.

Emboldened by the Voortrekkers' successes, other tribes opposed to Mzilikazi also joined the fight.

Attacked from all sides, in 1840 the ageing general led his people across the Limpopo River into what is now Zimbabwe. There he eventually established his capital at Bulawayo – 'the place of slaughter'. This was directly emulating his former hero Shaka, who also gave his royal kraal the same name.

Mzilikazi then organised his followers into a highly efficient military system with strategically placed regimental kraals and repelled the Voortrekkers. Unable to conquer the warrior-king, the Boers eventually signed a peace treaty in 1852.

Today the Ndebele people proliferate in western Zimbabwe. They are a living testament to Mzilikazi's incredible statesmanship, admittedly achieved at the sharp end of an *assegai*. Be that as it may, few, if any, other African leaders have been able to weld such a vast myriad of conquered peoples into a single tribe with such a strong and proud identity.

Mzilikazi was generally friendly to English-speaking whites, granting access to a varied assortment of hunters, naturalists, traders, evangelists and adventurers. Indeed, one of the most unusual stories of friendships to emerge from nineteenth-century South Africa was his relationship with the Scottish missionary Robert Moffat. This was the polar opposite to the modern-day story of antagonism between Kerneels and his neighbours in Mzilikazi's former home. It was a connection between black and white that could also be a parable for what we want to achieve at Loziba.

I know that's stretching it – but consider the background. It could not be more diverse. Moffat was pastoring with the Tswana-speaking people, who were on the receiving end of Mzilikazi's savagery. Seeing his flock suffer, Moffat was not a natural Mzilikazi ally. On the extreme contrary.

So there was little common ground for a friendship to flower. In fact, there was none.

Yet somehow the two men from vastly different cultures and beliefs formed an iron bond. Moffat described the king as 'charming, dignified, good-looking, with a ready smile'. He said that if he had not witnessed some of Mzilikazi's gorier executions first-hand, he would not have believed the monarch's fearsome reputation. For a pacific missionary to say that is mindboggling. Moffat was one of the bravest men alive at the time, believing that whatever he did was for the greater good of God. He was prepared to sacrifice his life for that. For Mzilikazi to have such a positive effect on him is something I believe goes way beyond the dry reportage of simplistic history.

Perhaps Moffat, like Mzilikazi, understood the deeper context of Africa at the time, something difficult to moralise about almost two centuries later.

The admiration was mutual. Mzilikazi called Moffat 'Mashobane', the name of his own father. For him, there was no higher accolade, and never could be. He also crowned him the 'King of Kuruman', the Tswana base of Moffat's mission. Any white trader or hunter who sought access to Mzilikazi had to enter Matabele territory on the Kuruman road, or they would not be received. And might be killed.

Most surprisingly, Moffat was allowed to say things to the king that would have resulted in a public execution for anyone else. The missionary regularly chided the warrior-king about his ungodly ways and gave him finger-wagging lectures on cruelty, which many would consider to be unimaginably reckless.

Yet Mzilikazi tolerated this with good humour. Perhaps even laughter. There was no chance in a million that he was going to change, despite his affection for this fearlessly devout white man. This was Africa in the nineteenth century. To hug an enemy was laughable at best, a death sentence at worst. The meek were the ones sprinting for their lives. They stood no chance of inheriting the earth.

After the Matabele headed north to evade attacks from all sides, the bizarre friendship continued with Moffat visiting his friend across the Limpopo. The years of battle and sweat, however, had taken their toll. Moffat recorded his shock at seeing the king's bloated body and palsied leg. He said that although Mzilikazi still invoked the deepest devotion of his subjects, he was no longer the 'Mighty Bull Elephant', the fearsome ruler of the past.

David Livingstone, Africa's greatest explorer, who married Moffat's eldest daughter Mary, also met Mzilikazi. He referred to the Matabele king as the second most impressive leader he encountered on the African continent. The first was possibly Chief Sechele, head of the Kwena people of Botswana, whom Livingstone managed to convert to Christianity. 'Convert' may be too strong a word, as Sechele's views on polygamy didn't quite chime with those of Livingstone. Sechele also believed that despite what the Bible said, his rain dancers were more effective in watering crops than going down on one's knees.

Mzilikazi died at his kraal, called Ingama, in 1868. That will be the name of the lodge once the Loziba Wildlife Reserve is established. A fitting name in honour of an absolute legend.

With the abundance of raw wilderness and rich history, Loziba was suddenly turning out to be one of the most intriguing conservation projects I had ever worked on. The Mzilikazi angle added a dynamic new edge that nobody outside the local tribes knew. The proposed reserve was living history where one of the continent's greatest warrior kings grew up, and where the rhino – *uBhejane* – an icon of the Zulu nation, would again roam free.

Our next goal was to get community support for the hugely ambitious Loziba project. There were several communities whose land was earmarked for inclusion, the most prominent being the Nhlazatshe, Hlonyane, Malangane, Nsabenkuluma and Tholulwazi clans. We had to speak directly to them all.

To do this effectively, I needed a Zulu talking to Zulus, so I asked my Rhino Art educator and partner Richard Mabanga to join me. Except no one in the outbacks of KwaZulu-Natal knows him either as Richard or Mabanga. He is *Mahlembehlembe* – 'all over the place'. Richard thrives in an exuberant life of semi-organised chaos and unlimited energy. As mentioned before, he always dresses in full warrior regalia for community meetings as he says this takes other delegates by surprise and they think he is 'important'. I couldn't disagree, as I always arrived in my *mblaselo* multi-patched Zulu trousers. We were a sight for sore eyes.

So with our maps and detailed blueprint, thanks to the most unlikely mastermind, Kerneels, the roadshow was now on. The stakes were high, kick-starting an entire region crippled with poverty and despair.

While the overall objective was the same, the meetings were vastly different. Some were in community halls dotted with a scattering of plastic chairs, others were in offices planned by bureaucrats with the architectural imagination of a rabbit designing a warren, or in the shade of trees under the blistering sub-tropical sun. All were lively, for the simple reason that this was happening in Mama Africa.

We never knew from one session to the next what to expect. Or what response we would get.

But none was more bizarre than a gathering where my main protagonist was a white professional hunter.

CHAPTER THIRTY:

The Will of the People

Africa is full of ironies, some bordering on parody. But this was perhaps the weirdest. It would be Kerneels's neighbours, the communities with whom he had spent decades quarrelling, who would see his conservation dream come true.

First, we had to convince them that it would not only benefit but transform the area, which was no small matter due to the legacy of bitter tensions. Also, just because we – Project Rhino-linked NGOs and Mawana Game Reserve – thought it was a good idea didn't make that a done deal. The most important stakeholders, the people actually living on the lands, had to buy into our vision.

I have dealt with many rural communities. They are not easily swayed. On the contrary, they're often extremely suspicious and with good reason. They have been victims of forked-tongue negotiations and pie-in-the-sky promises too often in their history. To convince them something will work needs bulletproof arguments. I believed that with Loziba, we had exactly that.

As we started the roadshows, sponsors came to our aid, which further convinced me that we were on the right track.

ERP funded the GPS collars for the runaway elephant herds and the ruinously expensive helicopter flying time. Roxane Losey, the Californian ex-jockey and a director of the Global Conservation Force who had cruelly (that's my opinion – she would say hilariously) highlighted my lack of saddle skills during that infamous anti-poaching patrol, once again sponsored equestrian equipment. We were also given four horses from Koubad, a nearby tourist farm lodge, that would be vital to halt poachers running riot on the Mawana reserve.

We also thought it would be a good idea to use the project as a test run for future community–wildlife research, so I contacted Dr Adrian Nel, a professor at the University of KwaZulu-Natal, for guidance. He teaches contemporary human–environment relations and suggested I take one of his Master's degree students, Phindile Mthembu, with us. It was excellent advice. Phindile would prove invaluable gathering information about the communities, their dreams and aspirations that we would otherwise not be aware of.

Our first roadshow was hosting members of the Committee for Rural Land Reform, as well as some headmen and chiefs of the larger communities. This was essential as Rural Land Reform would be one of three compulsory signatures needed on our land appropriation deals. The roadshow had to work, otherwise it would be back to head scratching for other strategies.

To my delight, I recognised the local Land Claims Commission director, Thulani Douglas Zungu. By extreme coincidence, he had been a teenager working for me on my former banana farm more than fifteen years ago. He greeted me warmly and said I taught him how to farm, which was very kind of him. This was a positive start. I could sense we would at least get a fair hearing.

However, that spark of optimism was soon doused. The rest of the committee was an hour and a half late and, when the chairman arrived, he seemed to be in a jovial mood. Convivial may be a better word. He was not in the least concerned that we were so far behind schedule that it was unlikely much work would be concluded that day. But this is Africa; you think on your feet and play the cards you are dealt. So we cut our losses and trudged up Mawana Mountain for lunch. As I plodded up the footpath, I remember Kingsley's pithy analogy of African punctuality: 'The Swiss invented time. But Africa owns it.'

Lunch on the mountain was a stroke of genius. Some of the chiefs had never set foot on the Mawana reserve before, due to the bad history. Yet here they were, honoured guests high-fiving each other and taking selfies at the edge of the rugged kopjes overlooking the primal valleys. Even more humbling was the positive reception that Beyers Coetzee, Kerneels's son-in-law, got with a PowerPoint presentation outlining our Loziba vision.

We then served up a cauldron of kudu stew, smoky and delicious from simmering on a log fire, while Beyers and I handed out frosted cans of beer or Coke. I could not have hoped for a better icebreaker.

Interestingly, there was also an impromptu debate between the headmen and committee members on how impoverished many people still were, despite some being granted former white land as far back as 2006. The benefits promised never materialised. Beyers and I looked at each other and nodded. This was exactly what the Loziba project aimed to rectify.

It seemed that the first hurdle, the chiefs and Committee for Rural Land Reform, had to some extent been cleared. Now we had to get the people on the ground to trust in the Loziba dream as well. This required us following strict protocol, and any deviation would likely lead to failure.

First, we had to obtain what was known as a 'Gatekeeper's Letter' from every chief representing every community. This would allow our Master's student Phindile to do socio-economic research, identifying community issues such as school attendance, cattle ownership, income and employment (if any), and also authorise us to speak to the various *izinduna* or headmen. The myriad *izinduna* are in Western terminology the community line-managers, who would then arrange for us to address the communities themselves.

In short, we could not speak to the *izinduna* without getting the green light from the chiefs, and we could not speak to the communities without the consent of the *izinduna*. There was no sidestepping that.

Fortunately, all were keen to hear what we had to say. One by one, the invites started coming in.

Our roadshows were usually held on weekends, as many community members are migrant workers from distant towns. Although they no longer live in their villages, they still have a say. They are also often the most educated and as a result, the most vocal.

This is not a bad thing. However, because they have jobs and food on the table, their prime concerns do not always chime with those below the breadline. As I was about to find out.

It was a day I remember well; a Sunday morning, with the sun blasting overhead, causing dark sweat patches to spread like saddlebags under the arms of the khaki shirt hanging loose over my *mblaselo* trousers. Squatting under the meagre shade of a paperbark acacia tree were sixty-eight people, waiting for me and Richard Mabanga. They were all members of the Hlonyane clan, one of the most destitute Mawana communities.

As we arrived, the *induna* – a Khumalo with the same surname as Mzilikazi – rushed up to the car and told us to keep driving as there was a 'problem'.

It turned out that a white hunter, Etienne van Wyk, who had a concession on the tribal lands, had arrived uninvited. Mr Khumalo said this could cause difficulties.

We drove further down the road until I got a call from the *induna* inviting us back. The hunter had left.

Mr Khumalo kicked off the meeting, introducing me by my Zulu name, *Nkunzi ayihlehli*. Basically, it means 'the bull never backs down' and was given to me by Zulu staff at the Shakaland cultural village started by Kingsley and Barry Leitch. I was 'awarded' it as the Zulus say I always finish what I start. No doubt the trick in getting such a name is to be careful what you start. But jokes aside, I am extremely lucky to have a Zulu name as my English one, Grant, is unfortunately synonymous with donating money – something conspicuously lacking in my bank account.

As mentioned, Richard is always introduced as *Mahlembehlembe*, so the roadshow 'stars' were a bull-not-backing-down dressed in bright Zulu pants, and an energetic guy in leopard skins. Say what you like, it was vastly different from any other land reform gatherings the community had attended.

I got up to speak, then noticed some commotion in the crowd. People were stirring uneasily. I looked over my shoulder. To my surprise, I saw Etienne the hunter and his manager had returned. They were standing about 20 yards away, arms folded and glaring at me.

The hair on the back of my neck rose. I sensed something was about to happen.

I didn't have to wait long. In fact, I hadn't even finished my sentence when Etienne started accusing me of being a troublemaker trying to build a park on his turf without consulting him.

I replied that we were not excluding him, but this was an introductory meeting with the landowners – the Hlonyane clan – not those, like him, who leased it. If there was a role for him to play, we would include him.

Etienne retorted that while he leased the land, he still owned all the animals and so had a vital stake in any future negotiations.

This was true. Even if the Hlonyane community terminated his lease, the animals would still belong to him. His family had owned the land before it was appropriated more than ten years earlier and he had lived there all his life. Just as the Van der Walts owned the game on Mawana, the animals on Etienne's leased concession were his property.

However, I tried to explain that this was not what the meeting was about. We first had to speak to the Hlonyane people. We could talk to him later.

Etienne was having none of that. He then yelled that it would be 'difficult' for him to work with the Van der Walt heirs as before Kerneels had died, he had laid eighty-seven criminal charges against Etienne. According to him, these ranged from attempted murder to poaching and theft. In his opinion, this was simply 'malicious'.

Wow – I took a deep breath. This could be a game changer. I knew that Kerneels had few, if any, friends in the area, and Etienne certainly wasn't one. But I didn't know the depth of that hatred had resulted in threats of violence and, according to Etienne, even allegations of attempted murder.

He was shouting at me in English, which I was translating into Zulu so the rest of the gathering could understand.

'Let's work together,' I said. 'We will never evict anyone from the park who has a right to be here. As soon as we do that we are going back to the old forced-removal days. It is not going to happen.'

Etienne shook his head, again shouting that we were trying to kick him off the land and destroy his business.

The meeting was getting out of hand. At one stage it seemed as though we might come to blows. He was shouting at me with clenched fists. I was preparing to defend myself.

Induna Khumalo stood, holding a hand in the air, commanding silence. The crowd quietened, including Etienne.

'This meeting is about the Hlonyane community. It is not about you two white men.'

I couldn't have agreed more.

He pointed at Etienne and his manager. 'Leave us.'

Etienne got up, saying he 'would never stand in the community's way'. I took that to be a positive ending to an otherwise belligerent encounter.

The *induna* looked at me. 'Continue,' he ordered.

I carried on with my speech, but this time instead of Etienne interrupting, there was a group of migrant workers seemingly more interested in what Richard and I were allegedly getting out of the deal than what the deal actually was.

The most prominent protagonist insisted on speaking in English. He knew I was fluent in Zulu, so that was either to prove his 'education' credentials or to highlight the fact that I was a stranger.

Either way it didn't work. Such divisive tactics rarely do in the long run. Instead, it gave me time to weigh my counter-arguments, as I had to translate all his queries back into Zulu for the benefit of most people present.

The thrust of his questioning was money. Firstly, where were we getting it from?

The answer was simple: investors interested in community development and conservation would be funding the project. We already had backers. All finances would be audited by accountants appointed by members of the community.

The next question was more pointed: what were we – Richard and I – making out of the deal?

Another simple answer. At the moment, nothing apart from some funds to cover our mounting travel expenses. There may be a lot of blood money in destroying wildlife – just ask the rhino-horn mafia – but not for those conserving it. John Kahekwa, living in a slum in Bakavu, is perhaps the most graphic example.

Another questioner asked about ancestral burial sites on Mawana where access was denied to the community. This was an important point, and I was grateful he brought it up. Honouring ancestors is central to many African beliefs throughout the continent, whether they be Christian, Muslim or animist.

'At Mawana we were loaded onto trucks and put in places where we didn't want to go,' he said. I could see many heads in the audience nodding.

I spread my arms towards the vast expanse of bush surrounding us. 'All of this will soon be community land. You will have unlimited access to all gravesites. Your ancestors will be proud that you have never given up. Loziba will belong to the community, and with that there will also be jobs and hope.'

'But if it is a game reserve for wild animals, how will we be allowed in?'

I responded, 'All decisions about Loziba will be made by the community. That will include the gravesites.'

Maybe my words fell on deaf ears. I do not know.

But the next speaker said words that I dared not. And they thundered around the valley. She spoke in colloquial Zulu, so my translation is not verbatim, and directly addressed the previous speakers.

'All these bad words about the new Loziba come from people who have jobs. We have nothing. So why don't you listen to this white man? Why don't you want to hear what he is saying?'

Then another man stood, his impassive face etched with hard living and his wiry beard flecked with grey. Again, I translate loosely.

'It's fine for you to talk,' he said pointing at the group of migrants. 'You are getting money every month, you are working in town. We who are here have nothing.

'Some of the Loziba land the white man is talking about was given to the Hlonyane people twelve harvests ago. We still have nothing to show for it. We are still hungry. Our children still walk far distances to school. Our women still fetch water. Even with the hunters coming and paying to shoot the animals, we have nothing. So let's listen to this white man and *Mahlembehlembe*.'

He sat down and there was significant applause. It was abundantly clear that many villagers were disappointed with the status quo. It was simply not viable. As John Kahekwa repeatedly said, 'Empty stomachs have no ears.'

Other speakers got up, listing a string of complaints about

their current plight. High on the grievance list was no running water, as the old man had said, few clinics, lack of public transport – and of course, they had no jobs.

There were other problems concerning the concept of a game reserve, such as they heard rumours that lions would be introduced. This was fiercely opposed as they said the big cats would kill their cattle. I made a mental note that we would have to address these concerns soon, as the vision for Loziba was that it would be a Big Five reserve. It was the only way to attract foreign tourism.

I pointed out that the entire reserve would be fenced, and all homes and cattle grazing areas would be protected.

This brought up a deeper issue. Many in the community were not convinced that a game reserve would bring them tangible benefits. One man stood, saying 'even the radio' believed that game reserves brought more hardships to the people.

Bizarre as that sounded, I knew exactly what he was referring to. A few months previously there had been a debate on Ukhozi FM, the local Zulu radio station, which had been an absolute public relations fiasco for the wildlife authorities. The programme was initially a discussion on rhino poaching, but soon developed into a disastrous free-for-all with the communities surrounding the Hluhluwe–iMfolozi Park venting their rage against Ezemvelo. Key resentments in a depressingly long litany of grievances had been that no compensation was paid for goats being eaten by escaped lions; low levels of employment from Ezemvelo KZN Wildlife who, they said, seldom recruited locally; and little direct community benefits from tourism.

It then got personal. One member of the audience showed his mangled arm that he said had been chewed off by a marauding leopard. Although his injuries obviously could not be seen by radio listeners, the reaction from the live audience confirmed it was true. He said he had been paid no compensation. Another man allegedly had been killed and eaten by a lion, whose family apparently also received no payment for their loss of a breadwinner.

That programme was several months old, and here Richard and I, standing under a gnarled acacia tree, were receiving the brunt of the backlash even though we had nothing to do with it. But the central question was something we had to address. If Ukhozi FM, the sole source of information for millions, was broadcasting news of communities not benefiting from wildlife, how could we counter it? What could we offer?

Fortunately, *Mahlembehlembe* had been a guest speaker on the Ukhozi debacle, and most of the crowd before us remembered him.

He stood, looking splendid in his tribal skins and headdress. 'The Hluhluwe–iMfolozi Park is owned by the government,' he said. 'Loziba will be owned by you, the people.'

He then gave examples of private parks that had enormously benefited the surrounding communities, such as Phinda, where the elephants in the valley several miles north had originally come from. Those animals one day would belong to the community, he said.

But, he stressed, most private parks are also not owned by the community, whereas Loziba will be. There will be jobs. There will be tourists paying a lot of money. And there will be compensation for stock losses, if any, as the people will decide.

Richard calmed the mood magnificently. We needed to get this positive message out loud and clear. The backlash he and I were getting from a radio broadcast from a different area concerning another community was worrying evidence of this. Ezemvelo was generally doing a good job in difficult circumstances, but we had to be absolutely clear that problems with government reserves would not be replicated in community conservation projects.

I also learned something else new that day. It came from a casual conversation when it was mentioned that there had been a number of contracts put out on Kerneels's life.

The man who told me this, a respected community elder, said that hired guns had tried to kill Kerneels on several occasions, hunting him as a sniper would. But the tough old Afrikaner was as wily as a Special Ops soldier, changing appointments at the last minute and often sleeping at different venues. It seemed he knew hitmen were out to get him, but always managed to stay one step ahead.

In fact, I later discovered that was exactly the case. Despite his unpopularity, Kerneels had a network of intensely loyal informers who worked under the cover of darkness. They were known as *uMagundwane* – rats – and fed Kerneels information about who was after him and when his life was in danger. Consequently, Kerneels knew all the hitmen's plans in advance. The word *uMagundwane* was also used to describe Kerneels himself, and the Zulu stuck.

This was sobering news – exacerbated by the claim that Kerneels had also laid an attempted murder charge against a neighbouring white hunter. Or so the hunter claimed.

I had to remember that in many cases this untamed land we were standing on was still an ancient frontier, albeit cerebral rather than geographical. People got killed over long-simmering feuds and quarrels, and here I was stepping into one of the biggest grudges in the district.

Was my life at risk? Was I in the crosshairs of assassins' rifles by those determined to make sure the Loziba project never saw the light of day? And that no deal be done with Kerneels's heirs?

Possibly not. But still, I was a ridiculously easy target. On most days of the week I would be driving alone in a tiny Toyota coupé on some of the remotest roads in the country.

There was not much I could do. On the plus side, it seemed that at the moment we were winning the community over.

But who knew what else was out there?

CHAPTER THIRTY-ONE:

Ghostriders in the Sky

I wasn't the only one being paranoid. Phindile also told me that she was increasingly anxious about visiting Mawana as she believed she could physically intuit the ghost of *uMagundwane* haunting the area.

She said the presence was so pervasive it was tangible, sending shivers down her spine. In her mind there was no doubt that the powerful spectre of the man loomed large over the landscape, despite the fact that he had died several months previously.

This came to a head one evening as Phindile was returning from doing field research with Mawana Game Reserve manager Paul van Deventer. Paul is a passionate young man of the bush, part of the new generation of committed conservationists, and acutely aware that the Loziba project has to succeed. He is doing good work bringing that about. A big plus is that he is popular in the area, mainly because he is unable to walk past a soccer game without joining in.

Paul drove Phindile to the lodge on Mawana where she would spend the night. She lives in Pietermaritzburg, about 220 miles away, and we would sometimes use the old hunting lodge as an overnight stop. It was furnished, but usually empty as there were no longer clients.

By the time Paul dropped her off the sky was darkening fast, so she used her cellphone torch to find her way around. Paul drove off to the manager's house about half a mile away. Both thought the lodge would be empty.

As luck would have it, Kerneels's youngest son Kallie had also been working in the area. It was getting late, he was hot and tired, so he had decided to sleep at the lodge rather than drive to his home in Vryheid close on an hour away.

He had brewed himself a cup of coffee, retired to the main bedroom and promptly fallen asleep.

With only the gleam of her phone guiding her, Phindile was unable to find the main light switch in the bedroom. So she sat on the bed to turn on one of the table lamps.

The bed heaved beneath her. It was a body. A face emerged – one that in the ethereal gloom of the twilight zone was the spitting image of Kerneels. A younger Kerneels. Phindile shot up and fled down the corridor, screaming at the top of her lungs, while Kallie rubbed his eyes wondering what weird dream he had been having.

Phindile laughed with the rest of us when she told the story. But I could see that nothing – nothing – would convince her this was not some eerie manifestation of Kerneels's ghost spooking the house.

She never stayed there again.

Meanwhile, I was also doing a bit of 'ghostbusting' myself, driving along the deserted roads, scanning the bushes to see if any hitmen were out to get me, imaginary or not. It was paranoia, of course, but strange things happened in that valley. For some, even Kallie's totally innocent but unscheduled appearance in Phindile's bed was somehow believed to be the old man having a hollow laugh from the grave.

However, there was nothing I could do about it. I had to travel on those desolate roads several times a week, speaking to the chiefs, headmen and communities. That was what the roadshow was about. The only alternative to my rising paranoia was giving up on the project, which was not going to happen.

What helped me most was that the tables were turning. And in our favour. Slowly but surely we started winning the communities over. More and more meetings produced positive results, but it was gruelling work. Often I had to drive a round trip of 400 miles from my home just to get a nod from an *induna* to arrange yet another meeting about a meeting. The time and petrol involved was crippling.

A normal day would start with me fetching Phindile at a taxi rank in Durban at 4 a.m. She would have left her Pietermaritzburg home an hour earlier. We would then drive to one of the villages surrounding Loziba to attend a community gathering scheduled for 10 a.m., arriving on time. The time slot was often ignored, with the meeting starting several hours later, but never earlier.

After that, I would drop Phindile off at a chief's kraal to get permission to do research in his area while Richard and I visited one of our Rhino Art school projects. Twenty minutes later I would get a call from Phindile asking us to come and fetch her. The chief wanted me and Richard to address the community, even though we had already done so. This could only be done in several weeks' time. Only then would Phindile be given permission to do research.

Finally, all the tangled jigsaw-puzzle pieces of the project started clicking into place. The picture we envisaged was starting to take shape. After close on a year of talks, travel, trials and tribulations, we at last had the verbal consent to proceed from most of the relevant communities, except the Hlonyane clan. That was the land where the big-game hunter Etienne van Wyk had a concession.

But even that was looking promising, despite the earlier bellicose encounter. We were talking to their leaders about paying a monthly rental while the project was evolving so they could see money physically coming in. We also pledged to employ up to fifty people as fencers and security guards, which would create more employment opportunities than the hunters were doing.

In fact, negotiations were so advanced that we started putting pegs in the ground where the 100-mile Loziba game fence would go up, keeping the animals off human habitation. The fence had been carefully mapped to go around every hut and village and it was important to do that now before the settlements expanded. Otherwise, we could face a similar situation as I had in the DRC where squatters were already inside the Parc de la N'Sele grounds by the time the fencing equipment arrived.

Our trump card was a raw nugget of priceless information I had picked up as an aside at one of the meetings with the Committee for Rural Land Reform. It was a point brought up privately when officials were asked if any appropriated land in the Vryheid and surrounding areas had been turned into something commercially viable. The answer was negative. The officials admitted that not one project, whether agricultural or conservation, had worked. Even worse, in most cases the people were poorer because the government's primary focus was on land redistribution, not skills and financial feasibility.

It was becoming increasingly clear that Loziba was being considered a blueprint for future development. Almost a lifeline. The chiefs, the headmen and the people on the ground were seeing first-hand that not much else was producing results. This eyewitness account of failure was far more powerful than bureaucratic promises emanating from urban offices far away. It could be the final roll of the dice.

Soon afterwards, the Loziba project got the most powerful ally it possibly could; a true legend in the often torturous story of modern South Africa.

But it took a tragedy – the largest single slaughter of rhino in KwaZulu-Natal – to bring us together.

CHAPTER THIRTY-TWO:

A Prince of the People

Gunfire shattered the silence. It was midnight, and Ezemvelo KZN Wildlife staff, rudely woken by the fusillade, said up to a dozen shots were fired.

Their reaction was immediate. Within minutes, rangers had grabbed their rifles and were calling on two-way radios, urgently asking for updates and situation reports. The question on everyone's lips was where the shots had come from.

No one was absolutely sure. In the still troposphere of the bush, the staccato crackle of gunfire could have erupted from anywhere. To pinpoint the exact area in pitch darkness was almost impossible. However, the general consensus was that the gunmen were somewhere in the Mbhuzane sector.

Nights in the Hluhluwe–iMfolozi Park are regularly punctuated by gunfire as the rhino wars rage. But these are usually muffled blasts, a rhino shot by Mozambicans or gangs from Mpumalanga. If there was a calf, the poachers invariably killed it with an axe blow to save bullets. So one shot could equal two deaths.

This was different. A volley of shots, meaning the killers were armed with assault rifles, probably AK-47s, and involved several rhinos. Something dreadful had happened; the rangers felt it in their bones.

They scanned the bush for flash or vehicle lights. Nothing. The night was quiet once more. The cicadas tentatively resumed their hum, the ubiquitous background harmony of the bush.

There wasn't anything the rangers could do. If they knew where the shots had come from, they would have set off instantly. This may not always be advisable in the night, as in Africa darkness favours the predators. But there would be nothing holding the rangers back if they knew where the men with guns were.

They sat waiting for the sun, cursing the minutes ticking agonisingly by. Each second would make their job more difficult. The poachers would most likely be miles away by the time the chase was on. The fact that the killers had brazenly fired multiple shots meant they knew the rangers would never catch them.

As the first rays glittered pink-gold on the eastern horizon, the Ezemvelo men clambered into the back of a 4 x 4 bakkie. Black specks circling high in the brightening sky told the story as clearly as a written note. The vultures were already out, swooping down in spirals to feed. The huge number of birds indicated the feast was of gargantuan proportions.

The men used the vultures as they would a point of a compass. Their initial instincts had been correct. The gunshots had come from the Mbhuzane sector, an area known for its magnificent wilderness trails.

Forty minutes later they found what they were looking for. It was their worst nightmare.

Close to the banks of the majestically wild Black iMfolozi River were six dead rhinos. They lay in crimson pools of congealing blood. All had been messily dehorned with axes. Their mutilated faces, matted with buzzing blue-bellied flies, reflected the silent horror of human avarice.

The rangers stared and seethed, fists clasping and unclasping in impotent fury. Then they sprang into action, following tracks of boots and blood at an almost reckless sprint, such was their rage. They wanted revenge for these cruelly murdered animals who could not avenge themselves.

But the killers had a running start of close on six hours. The tracks led to the fence, then to a road. Tyre marks told the rest of the story.

This was, and still is, the worst single incident of poaching in Ezemvelo KNZ Wildlife's history.

It also signalled a frightening spike in the rhino wars. A multimillion-rand, super-sophisticated surveillance system installed in the Kruger Park was providing spectacular successes with scores of poachers arrested – but conversely, this breakthrough added another horrific dimension to the war. It viciously kicked open a new front as Mozambican and Mpumalangan gunmen working for the various horn syndicates started looping south into KwaZulu-Natal, where they considered the pickings to be easier.

This is not to say that there had been no horn poaching in KwaZulu-Natal beforehand. On the contrary – but it had never exploded on the scale we were now witnessing.

To make matters worse, wildlife authorities had been warned this blood-horn tsunami was surging our way. Local communities, often at great personal risk, told Ezemvelo that *kwerekwere* – foreigners – speaking Swati and Shangaan-Tsonga, northern Nguni dialects similar to Zulu, were coming into their villages.

The danger signs were in plain sight, and had been so for more than a year. Despite this, the slaughter continued unabated. The murder of six rhinos in one hit was the culmination.

To add bitter irony, it happened on 22 September 2016. It was World Rhino Day, a global awareness event set aside to highlight the plight of the world's most threatened creature. It was also exactly two years to the day that the first World Youth Rhino Summit had been held in the same park.

Equally poignant, Mbhuzane in the south-western sector of the park was the cradle of Operation Rhino, where six decades ago the southern white rhino was first rescued from global extinction.

The latest butchery raised the annual rhino body count in KwaZulu-Natal to 139, suggesting that the final death toll in the province would exceed 260 by the end of the year. In other words, three rhinos killed every two days.

It would be a record soaked in blood.

I had not felt so outraged since the saga of Thandi, the young rhino savagely dehorned with machetes in the Kariega reserve four years ago. Thandi survived, thanks to ground-breaking veterinary work by my brother William and his team. These six animals stood no chance.

There was nothing we as NGO operators could physically do as the animals had been shot on a state reserve. On a private reserve such as Amakhala, staff track each endangered animal around the clock. We have roll calls. We know where every rhino or elephant is every minute of the day. State reserves, with much larger animal populations, don't have that safeguard. However, this massacre certainly sped up my plans to help Ezemvelo with establishing anti-poaching horse patrols. Less than a year later, Mbhuzane, the scene of the tragedy, became the main base where the horses were stabled.

The carnage continued. Next was an attack at the Ophathe Game Reserve near Ulundi. The park, also run by Ezemvelo, was unusual in that all the resident rhino had been surgically dehorned to deprive poachers of their grisly prize.

Well . . . almost all. One female that had recently given birth had been excluded as veterinarians would have had to dart the baby as well and did not want to traumatise mother and calf. She was scheduled to be dehorned the next time the vets visited.

Poachers came that night. They killed the mother and hewed off her horn. Then they killed the baby, also hacking out the tiny stub of her immature protuberance.

I flipped my lid when I heard this.

Another particularly brutal incident that affected me happened at the Kragga Kamma Game Park near Port Elizabeth, where a rhino cow had been surgically dehorned five days beforehand by my brother.

It didn't matter. The poachers shot and hacked her to death anyway. Her fifteen-month-old baby fortunately escaped, unlike the infant at the Ophathe reserve.

This begs the question: are these men so evil that they will callously kill a rhino that is of little use to them? Just for the hell of it?

The answer is probably that poachers hunt at night, so are unable to see if a rhino has a horn unless it is directly silhouetted. However, some conservationists believe that the poachers kill a dehorned rhino for a supremely callous reason: to prevent wasting time tracking it again. This is what we are up against.

Reserves are now putting up large signs stating that their rhinos have been dehorned, and we at Rhino Art are also telling the schoolkids that, hoping the bush grapevine information reaches the poachers.

The next heartbreaking incident happened at the Wildschutsberg Game Reserve near Tarkstad in the Eastern Cape. In this case, the reserve's entire rhino population was wiped out.

Once again, the rhinos had been previously dehorned, so the poachers escaped with bloody stubs worth a fraction of a full horn. One of the five animals initially survived the mutilation of its face, but died a few days later. Greg Harvey, owner of the reserve, summed up the situation in a newspaper interview: 'It is just madness. I had five rhinos that were of breeding age and now they are gone in the blink of an eye.'

Greg's simple words were a chilling reminder of what was at stake. Gone in a blink of an eye. Time was certainly not on our side.

However, the most brutal attack was at an orphanage for baby rhinos as this time the victims were not only animals but young volunteers from Europe. This resulted in global publicity, coupled with the fact that it happened at Thula Thula, one of South Africa's better-known reserves. This was where internationally acclaimed conservationist Lawrence Anthony had rescued a rogue herd of elephants, a story made famous in the bestselling book *The Elephant Whisperer*.

The reserve is now managed by Anthony's indomitable widow Françoise Malby-Anthony. However, her team were not responsible for managing the rhino orphanage or the security arrangements. This was handled by a conservation charity that works independently of Thula Thula staff.

The poachers stormed the orphanage, ambushing the security guard, ripping his firearm off him and disabling the CCTV cameras.

Then the orphanage staff – mainly young women – were violently beaten as the poachers shot two rhino calves, Impi and Gugu, and hacked off their tiny horns.

The volunteers later needed trauma counselling. It's difficult to imagine their terror – First World idealists barely out of their teens in love with nature, working in a foreign land, being viciously assaulted and seeing the animals they cared for being butchered like slabs of meat.

I started to despair. The evil we were confronting seemed so overwhelming.

There had to be something I could do. My expertise was not running around the bush with a gun. Instead, it was shining the harsh glare of publicity on these atrocities, shaming a lethargic world and lacklustre politicians into action. This was something the wildlife authorities were not particularly good at. For example, the World Rhino Day massacre at Hluhluwe–iMfolozi received scant media coverage, just a couple of stories stating that the killings had happened and some statistics of rhino mortalities. The tragedy, the emotion, the cold-blooded barbarism, the sheer satanic evil was not mentioned. Until William took a video of Geza, the rhino born on Amakhala who died in the most agonising way possible, most people thought poaching was a relatively bloodless affair. A gun was fired, an animal fell. Geza, followed by Thandi and Themba, brought to the world in full technicolour the gruesome reality of the horrific suffering involved.

I was lying in bed, swamped with despondency, when it came to me. I stared at the ceiling and decided, cold as a snake, that I would take these atrocities as far up the political ladder as I could. I would make sure it got the attention of the highest office of the land – Parliament itself. I would not rest until that had been done.

The two most influential people in the Zulu nation are the monarch, King Goodwill Zwelithini, and the leader of the Inkatha Freedom Party, Prince Mangosuthu Buthelezi. Although the Zulus were seldom rhino poachers themselves, they sometimes harboured the gunmen in their communities out of fear. If we could get the king or Prince Buthelezi, who had been a cabinet minister under Nelson Mandela, to vehemently denounce the wholesale slaughter of rhinos with a powerful statement, it might galvanise the masses.

King Zwelithini demonstrated the power of his words the year before, although not in the way he intended. He said in a speech that 'foreigners should leave South Africa', which unwittingly sparked a wave of xenophobic violence. He was referring to the millions of migrants pouring across the borders and allegedly taking jobs from locals. It was certainly not intended to be a call to arms – the king was cleared of that by the Human Rights Commission – but unfortunately many of his subjects took it that way.

That had been unfortunate, to put it blandly. But if the king could provoke reaction from something so negative, imagine the possible reaction from a supremely positive message?

However, Prince Buthelezi was my first choice as he is an inspirational political leader who, like me, also loves rhinos.

But how to get to him? It's not a case of picking up the phone and saying, 'Hey, let's have a chat about this pachyderm problem.'

I contacted an old friend, wildlife expert Paul Dutton, and

somewhat brazenly asked if he could arrange a meeting with the legendary leader.

Paul had been a game ranger in the days when the KwaZulu-Natal nature administration was known as the Natal Parks Board, one of the most revered wildlife authorities in the world. He's lived a fascinating life, doing conservation work throughout southern Africa, and was once jailed by the FRELIMO government during the civil war in Mozambique. More importantly, as far as I was concerned, he had worked with Ian Player and his mentor Magqubu Ntombela, and thus had a direct personal connection with Prince Buthelezi.

But even so, I thought this whole idea was at best a long shot in the dark. So imagine my astonishment when Paul called the next day to say we had been invited to Prince Buthelezi's office in the Zulu capital of Ulundi for lunch.

To give a brief – most would say too brief – background, Prince Buthelezi is a political giant, as highly revered as Shaka or Mzilikazi in much of Zululand. A fierce opponent of apartheid, he fought the system tooth and nail from within, never going into exile, and always insisting on non-violent opposition. This may not have won him accolades from the firebrands, but it is no secret that it is people such as him, true statesmen, who stopped this country going up in flames during the dying days of apartheid.

He is also of royal blood, a prince in Western terms. He has ice and fire in his veins. The fire of a warrior, and the ice of intellect and reason. He is the great-grandson of King Cetshwayo kaMpande, the Zulu king during the Anglo-Zulu war of 1879. Cetshwayo was the last monarch of the old order, when Zululand was a single kingdom. He led his people to victory in the Battle of Isandlwana, the most devastating defeat ever inflicted on an imperial army by warriors armed with spears and a handful of rusty Martini–Henry rifles. Interestingly, in the blockbuster movie *Zulu*, Prince Buthelezi played the role of his maternal great-grandfather, which was a nice touch in a film that honoured the exceptional valour of both Zulu *impi* and British soldier.

We were told we could bring a delegation of six people, and Paul and I spent some time choosing whom we thought would best fit the bill. Richard *Mahlembehlembe* Mabanga and Kingsley Holgate were my first choices, along with our new Project Rhino head Chris Galliers. Paul then chose the legendary wilderness guide Mdiceni Gumede, and we both agreed to include one of the country's top new generation of conservationists, Musa Mbatha, the assistant manager of Phinda Game Reserve.

Unfortunately, Kingsley was wandering around somewhere wild in Somalia at the time on another of his epic expeditions, so he missed this opportunity with huge regret.

What followed was possibly the most humbling session of my life, spending several absorbing hours with an icon who had spent most of his life fighting the apartheid regime.

Prince Buthelezi was eighty-nine years old at the time, but you would never have guessed that from his impressive energy and vigour.

It was a jewel of a day. The ninety minutes we had been allocated in the Zulu leader's hectic schedule turned into four hours. We laughed at old wilderness stories, wistfully recounted by Mdiceni, who told of walking with wild rhinos, drinking in the rivers with them, and sleeping under the stars with only these prehistoric pachyderms for company.

In those days, rhinos had the God-given right of dying from

old age with dignity. For some reason, their horns barely decomposed and it was not unusual for a ranger to use an old proboscis found in the bush as a door stopper. They had no cash value. There was no insatiable human greed. At least not in that innocent time in the outbacks of the Hluhluwe–iMfolozi wilderness.

'I have spent sixty years in the bush,' said Mdiceni. 'Nature is my mother. We suckle at her breast. But now I have seen these people – the poachers and their bosses, killers from distant lands – steal our heritage, our gift from God.'

His sorrow was so primal that it was palpable. Prince Buthelezi wept.

We then visited Prince Buthelezi's personal museum. On display were relics of the apartheid era he had fought so valiantly to abolish. He also has a fine collection of Zululand memorabilia. Among his mementoes was a wonderful black-and-white photo of him and Ian Player, taken during the heyday of Operation Rhino. They were smiling at the camera, both strong, fit young men, committed to what they knew to be right. Prince Buthelezi was a guiding light in getting the iMfolozi Corridor incorporated into the reserve, turning it into the modern flagship biosphere of the Hluhluwe–iMfolozi Park. This country, both its people and its animals, owes him an enormous debt.

Getting back to the rhino crisis, he asked us to draw up a plan of action that he pledged would be tabled in Parliament.

This we did and emailed it to him. True to his word, a ten-point memorandum that we drafted was read out by the Speaker of the Inkatha Freedom Party, Narend Singh, on 13 June 2017.

Among the key issues were:

- The Treasury Department must count the cost of illegal rhino-horn trafficking.

- Tourism must count the future cost of lost revenue.

- The Department of Justice must create circuit courts in all districts where protected areas occur.

- The Police must investigate the syndicates who are operating freely.

- The Department of Safety and Security must refit the military with green berets to defend the country's natural heritage.

- The Department of Environmental Affairs must study the pros and cons of dealing with existing stockpiles of legal horn.

- Communities living around protected areas must receive incentives and training to become partners in this fight.

Basically, it boiled down to political will. Bringing in the military to help fight the bush wars instead of relying on out-gunned rangers; creating circuit courts for swift justice; and empowering local communities living near nature reserves to join in the fight.

Sure, this would need funds, but as the memorandum stated, the country was losing far more money in illegal horn trafficking and potential loss of tourism than it would in facing up to the crisis.

The day the memorandum was read out was a significant moment in the rhino wars. It was when proposed solutions were actually documented in a parliamentary session. While we were not that naive to think the government would react instantly, if at all, at the very least a working document to deal with the catastrophe had been tabled in Parliament. The ball was in the court of the highest office in the land. In a small way, history had been made.

I didn't realise it at the time, but that lunch at Ulundi had far-reaching implications for me. Prince Buthelezi is a patron of the Wildlands Conservation Trust and consequently extremely interested in the conservation work we were doing with Rhino Art and Project Rhino. I was now more likely to get access to his office, without having to rely on my friend Paul Dutton.

This led me on to another meeting with Prince Buthelezi, in which I cautiously broached the subject of the proposed Loziba project. Loziba is directly linked to rhino conservation as one of the top priorities is to restock the land with black and white rhinos. As a result, it was a natural progression of our conversation.

I outlined the project, highlighting our eco-vision and the historical connection to Mzilikazi. I could see I had his interest right away. Although Prince Buthelezi's royal bloodline flows from the house of King Shaka and Mzilikazi was considered guilty of treason, the proposed Loziba Wildlife Reserve was living, breathing history.

Obviously, I didn't expect any public endorsement while we were still talking to their various communities as that could politicise the matter. There was no way I would do that, nor would Prince Buthelezi interfere in what was essentially a community land issue.

However, he asked me to keep him updated. I said I would

once the project was up and running and all communities had signed. When that happened, I knew we had an extremely powerful ally on our side.

Meanwhile, we got a boost from another royal connection. This time from the man fifth in line to the British throne.

CHAPTER THIRTY-THREE:

When Harry Met Bill

As anyone who knows my father will attest, he enjoys whisky, tobacco and colourful language in equal proportions. Equally large, that is.

So he was bemused when William came up to him saying a special visitor was arriving the next day, and he would have to limit himself to two small whiskies, no pipe smoking and zero swearing.

My dad, whom everyone calls Uncle Bill or Tick Bird, was having none of that. No guest was 'that bloody important' for him to have to resort to such drastic measures, he said. Most people would have called the 'drastic measures' mere good behaviour.

'Just for two nights. Please, Dad,' said William.

The next evening Bill was sipping his first Scotch when a familiar figure walked in.

It was Prince Harry.

Bill, sitting at the bar counter, squinted in the gloom of the impressively atmospheric stone-wall room adorned with wild-life artefacts, which was once a cellar more than a century ago.

'Haven't I seen you somewhere before?' he asked.

Harry laughed.

'What do I call you?' Ticks asked.

'Harry.'

'I'm Bill. Welcome to my pub.'

In that instant, all airs and graces were not just thrown out of the window – they were hurled. Harry was in for a convivial ride on the wild side by probably one of the crazier people he has met in his life.

His visit was so secret that even William, his host for the next two nights, was only informed His Royal Highness would be coming to Amakhala forty-eight hours beforehand. Even then, he couldn't tell anyone else until Prince Harry actually set foot on the property. With some dread, he realised the key issue was getting Bill to behave. But he figured that as Harry had been to Africa on several occasions, he would know strange things happen on this continent.

Harry's visit was primarily to experience work as a wildlife volunteer and witness first-hand the rampant rhino- and elephant-poaching crisis. The trip started in Namibia, and then he flew to the Eastern Cape with a group called Saving the Survivors to watch an operation on a rhino that had survived a particularly vicious poaching attack. From there he would visit KwaZulu-Natal, Botswana and Tanzania.

The mutilated rhino's name was Hope. It had been darted and dehorned with machetes while still alive at Lombardini, a wildlife park near the world-famous surfing mecca of Jeffreys Bay. That the rhino had been felled by a highly potent drug only available to veterinarians was another growing concern, indicating a serious breach in the medical supply chain. Using a tranquilliser means no shot is heard by game guards, and the horn is hacked off while the animal is alive but paralysed.

Hope was first treated at Lombardini by William and Johan Joubert, and then moved for further intensive treatment to the Shamwari Wildlife Rehabilitation Centre, where Johan is the resident veterinarian. That's where Harry would watch my brother and Johan in action.

The day before the Prince arrived, his personal security guards pitched up at Amakhala and outlined the protocol that must be strictly followed. All guests had to be cleared out of the premises and given alternative accommodation on the reserve, and all cellphones had to be confiscated. A big concern was that a candid photo of the Prince would end up on Instagram or WhatsApp.

Above all, no one was to be told of the visit. Sadly, that included me.

Both William and Johan do a lot of work for Saving the Survivors, treating rhinos with gunshot wounds, facial gouges and other poaching injuries. The charity has so far treated ten horribly injured rhinos in the last few years, of which nine survived. It's an incredible record. However, Hope's wound, measuring 20 inches by 11, was among the worst that William and Johan had seen.

Harry travelled with William in the front of the Land Rover to the rehabilitation centre. There he took one look at Hope's grotesquely mangled face and said the vets were wasting their time.

'Isn't it better just to take a gun and shoot her?'

William shook his head. 'What I want to show you later will hopefully change your mind.'

Harry was going to see Thandi, the first ever rhino-poaching survivor, and her new calf that afternoon.

'OK. I'll reserve my opinion.'

Harry watched as William and Johan worked on Hope. Although the rhino had a face mask crafted from a skin graft, the nasal cavities and some other vital skull organs were visible.

That was the 'before' scenario. That afternoon, he saw the 'after' with Thandi at the Kariega reserve. The Prince was both astonished and visibly moved. Although Thandi's scar was noticeable and she had no horn, she was just like any other healthy rhino.

Harry looked on, fascinated as the rhino that had defied certain death contentedly fed her new calf Colin, named after Colin Rushmere, the founder of the Kariega reserve. Colin's son Mark Rushmere, a former South African test cricketer, now runs Kariega with the Fuller family.

'It's incredible,' Prince Harry said, retracting his earlier comment. 'I now know what you mean. I see what you have achieved.'

His voice was quiet with respect. Watching Thandi and her calf grazing peacefully in the savannah said far more than words ever could.

From there, they went to Amakhala. Harry and his entourage would be spending the next two nights at the Leeuwenbosch Country House, built in 1908 on land that has been owned by the Fowlds family for five generations.

As the sun set, the royal guest popped into the stone-walled bar for something long and cold and to meet my dad and Rose, my mom.

There were no airs and graces. It was like having a drink with friends at the local tavern. On our side were my parents, William, his wife Heidi, their children, and Heidi's mother Natalie.

With Harry were his two bodyguards, who basically had nothing to do as I don't think they had ever been to a more peaceful place than Leeuwenbosch. They didn't even have to look under the beds – in fact, the only peril was politically incorrect comments from my dad.

Also there was Harry's close friend, Inge Solheim. My family held their collective breaths as the Swede was introduced to Bill.

'This is my friend Inge.'

Bill looked him up and down.

'But that's a girl's name. I once had a girlfriend called Inge.'

There was a second's silence, then a guffaw. It came from Harry.

'I promise you, Bill, this one's got balls.'

That was no lie. We later discovered that Inge Solheim is a hardened polar adventurer described by *GQ* magazine as an 'extreme nomadic philosopher' making Britain's best-known outdoorsman Bear Grylls look like a 'toddler playing in the sandpit'. He is also an integral part of Harry's 'Walking with the Wounded' charity, rehabilitating injured army veterans.

That was the start of it all. The banter and laughter went on for several hours as everyone except the kids and bodyguards smashed through the minimal alcohol barrier and colourful language ban imposed by William.

On the second night, Harry said he had enough of beer and wanted to have a whisky with Bill. The barman served up two industrial-sized J&Bs.

Bill then asked Harry why he wasn't married. He supposedly had every girl in the world after him.

This was before Harry had met Meghan Markle, so he hit the ball squarely back into Bill's court.

'Good question. Can you find me a wife?'

Bill mulled this over for a moment, then replied that the task would be infinitely easier if Harry was better-looking.

Once again, the Fowlds family members held their breath – mortified. Once again, the loudest hoot of laughter came from Harry.

Bill was in fine form, recounting his famous story of when he was in hospital after a knee operation and had to go to the toilet on a wheelchair with a hole in the seat to provide access to a commode.

However, without providing too much information, suffice to say that my dad refuses to go to the toilet without lighting up his pipe at the same time. This was a non-smoking zone, which meant nothing to Bill, but another person in the toilet raised the fire alarm thinking the impressive clouds of smoke indicated a blaze had broken out.

The nurse on duty, Sister Iverson, rushed down the passage in full fire-fighting mode, opened Bill's toilet door, grabbed the wheelchair as she could not see my dad in the billowing tobacco smoke, and in the resulting wild melee jammed his nether regions between the porcelain commode and the chair. She pulled so hard that one wheel was spinning in the air while my dad was yelping in agony. She then called another nurse to help, and between the two of them yanking and tugging, they somehow freed my dad's wedding tackle which then plopped into the commode.

Harry almost collapsed he was laughing so much.

Despite Bill being as tough as old boots, he is extremely sentimental. I think that's why so many people love him. True to form, after dinner he gave an impromptu speech about Harry being the first British royal to visit Amakhala, and what an honour it was. He said he was like his mother Princess Diana with his tireless work for good causes. She really cared for people, as demonstrated with her charities ranging from clearing landmines in warzones to AIDS and leprosy.

Harry put up his hand and asked Dad to stop. Tears were rolling down his face.

This got my dad going as well. There they were, a prince and a weather-beaten old man of the bush, weeping together.

There is no doubt that my crusty old dad and Harry hit it off. We later discovered that Bill's fame – or more likely notoriety – had reached the rest of the royal family. As fate would have it, my brother William later met Prince William at the Tusk Conservation Awards and he mentioned that the Fowlds family had recently hosted Harry at Amakhala.

Prince William nodded with a knowing grin. 'Harry tells me your father is an absolute legend. I would love to meet him one day.'

So obviously Harry had spoken to his brother about his two nights of drinking in a bush bar with a crazy old man.

I arrived at Amakhala the day after Harry left, oblivious that he had been there due to the shroud of secrecy my family had sworn to uphold. My mother was particularly devastated as she said Harry and I would have had a lot to talk about as he is also interested in promoting conservation among youngsters. He would have been fascinated to hear first-hand about Rhino Art, she said.

Then to add to my all-round bad timing, Harry visited ZAP-Wing, the flying arm of Project Rhino in KwaZulu-Natal, when I was away. He was taken on a flight over the Zululand reserves and, again, the visit was cloaked in such secrecy that I only heard about it afterwards. In fact, all publicity about

Harry's African jaunt, mainly Instagram photos and his personal blog, was only released several months later when the Prince returned to England.

Sadly, not too long after Harry left, the rhino Hope died. She had been well on the way to full recovery and everyone was optimistic a miracle would happen, as it had with Thandi.

She was then moved to the Onderstepoort Faculty of Veterinary Science in Pretoria for further treatment when she suddenly died from an additional stress-related infection. Yet another horrible statistic in the poaching crisis.

But Hope had served her species well. Her valiant fight for life, as well as being introduced to Prince Harry, touched the hearts of thousands of people. Like Thandi, she was a rallying cry in the wildlife wars.

However, there was a glimmer of good news elsewhere. The two key projects I had been working on started coming to fruition. Slowly, bit by bit, the wheels of progress started turning.

I also began to clarify my own vision of the future for our persecuted wildlife.

CHAPTER THIRTY-FOUR:

Visions of the Future

It always was a big ask, and many thought my plans for mass elephant relocation from the Atherstone Collaborative Nature Reserve were too ambitious.

To find a home for eighty Limpopo province tuskers on other South African private reserves was expecting too much, particularly as we were reluctant to separate families. This meant that if one reserve could only take six, we could not send it a family nucleus with more than six members. We could not customise orders.

I started thumbing through my cellphone contacts, phoning reserves that might be interested. Most were regretful refusals.

Then I hit the jackpot with two takers; the Buffalo Kloof Private Game Reserve near Grahamstown belonging to my cousin Warne Rippon, and Mount Camdeboo near Graaff-Reinet in the Great Karoo. Both only wanted a small number – eighteen in total – but it was a start.

It was like Arnold Schwarzenegger himself had bench-pressed the weight off my shoulders. Obviously the deals would be subject to permits and the reserves getting Elephant Management Plans approved, but the main thing was that in a reasonably short space of time I had tentatively sourced magnificent homes for close on a quarter of the elephants that needed to be rehomed.

However, my sense of urgency in getting the animals to their new ranges was not necessarily shared by the wildlife authorities. For example, just to move the animals required a TOPS (Threatened or Protected Species) transport permit, and that alone took me almost three months. Then when it was signed, the date was wrong and I wasted another twenty-four hours rectifying that. I was convinced that while *uMagundwane*'s ghost was apparently haunting Phindile at Loziba, the pseudonymous spook Phillip Hathaway was doing likewise to me at Atherstone.

I also soon discovered that despite my initial success, most of the other private reserves are at saturation point with elephant populations. Elephants may be endangered, or even extinct in their historic home ranges throughout much of Africa, but that is not the case in South Africa. They are flourishing on private parks. This is despite the fact that they are not easy animals to accommodate for the obvious reason that they are so large. They devour several hundred pounds of foliage a day. Small reserves are soon depleted of even fast-growing red grass and shrubs. Also, a savannah elephant's propensity to shoulder down trees, to get at either top leaves or roots, means decent-size woodland is required.

The obvious solution is to move them to other state parks, which are usually substantially larger than private reserves. But that is a paper solution only found on some city-slicker bureaucrat's desk. The reality in most cases is that it's a non-starter. Sadly, many government parks are running on empty. While they may have the space, most cannot take more elephants due to lack of funds coupled with management problems. That's the situation on the ground and there is no way of delicately skirting around it, or pretending it doesn't exist. To place elephants in a significant number of state-run reserves would pose a breakout risk to surrounding communities, possibly life-threatening, and an increase in ivory poaching due to insufficient security.

The Atherstone situation also highlighted other far-reaching demographic problems. We had to curtail the rocketing elephant population in ever-decreasing wild areas. The only way to do that without culling was implementing birth-control programmes. However, the jury is still out on how effective immuno-contraceptive injections, which make females temporarily sterile, actually are. The outcome of this is critical for future wildlife planning. If contraception doesn't work at least most of the time, there would be no other option but to move South African elephants to parks in other countries. Or culling, but no one wanted to verbalise that road-to-hell route. There was too much at stake to consider that – and certainly not on the watch of people like me.

So, to add to my woes, I realised that to find new homes for the remaining sixty-two of Atherstone's elephants could mean looking elsewhere on the continent.

This would obviously cost truckloads of money. Transporting 6-ton creatures thousands of miles is a highly specialised and expensive undertaking.

This begged the next question: where would I get those funds? The answer, as always, is from NGO organisations.

However, many NGOs are suffering from donor fatigue and so finance needed for such an intensive logistical operation was certainly not guaranteed. We were all already operating on a shoestring.

I suddenly discovered that the Atherstone situation I was working on was just the tip of a nightmarish iceberg, highlighting a host of other problems piling up through increasing human encroachment.

On the other hand, it might force us to adopt other out-of-the-box answers.

I then discovered that I was already sitting on one.

Loziba.

While Loziba could not take more elephants because of the three resident herds already there, it was an example of thinking creatively as the wave of land appropriations swept the country. As mentioned earlier, much of the agricultural and wild land available was going to communities who had little idea what to do with it, and the answer to that problem is conservation – particularly Big Five reserves. That's exactly what we were planning for Loziba. Just as Atherstone could be a blueprint for re-homing large numbers of elephants, Loziba could be a beacon for other community-owned conservation projects accommodating both surplus and endangered animals.

Those were my uppermost thoughts on a quiet Sunday evening at a village on the border of Mawana Game Reserve. The sun was setting behind a patchwork of cotton-wool clouds, igniting them blood-red for several glorious minutes. In front of me were representatives of the Malangane clan, the largest of the various groups interested in the Loziba development. We were there for a historic meeting.

The community had agreed to endorse the project, which was a giant step forward. Not only that, we at last had it in writing. In a letter signed by the Director of the Land Claims Commission, Thulani Douglas Zungu, the following declaration was tabled: 'It was agreed that the proposal was an ideal project for tourism, job creation, skills development and opportunities for an area that had not seen the full benefit of Land Redistribution. It was further agreed to endorse this project and endeavour to fast-track the outstanding Land Claim sales and portions that were claimed by the various Communities being Hlonyane, Malangane, Geluckstadt and Nhlaztshe claim.'

This was the consent we needed, the final step before proceeding to a signed contract.

It was not an actual go-ahead. To paraphrase Winston Churchill, it was not the beginning of the end, but the end of the beginning. We now had the overwhelming majority of the communities on board. And, of course, with Dr Mangosuthu Buthelezi tentatively in the loop, this had enormous potential for the future.

However, land expropriation issues can take years to untangle. As promising as Loziba seemed, we needed more immediate solutions for situations such as Atherstone.

The answer lay in the rest of Africa. But as soon as I realised that, another question rattled in my head: where exactly in Africa?

On our borders, Namibia and Botswana, both with excellent conservation credentials, can't take more elephants, and in Zimbabwe the two most famous national parks, Hwange and Gonarezhou, are also saturated.

In Malawi, the biggest elephant relocation in history had recently taken place with 500 elephants being moved over 200 miles from the Liwonde National Park and Majete Wildlife Reserve to Nkhotakota Wildlife Reserve, thanks to the poaching problem on the Mozambique border being resolved. In fact, Prince Harry had been at that mass relocation helping as a wildlife volunteer soon after leaving us at Amakhala. So Malawi would also be unlikely to want more elephants.

This sounds as though the problem is elephant over-population rather than conservation. That is not the case. It is human encroachment. We are trying to downsize elephant herds because we are taking away their historic ranges. It's as simple as that. Once that fact is grasped, then the answers become clearer. We have to find wild land with minimal human encroachment. Land where they can roam as life, as they know it, intended.

The good news is that there is plenty of it. The bad news is that we cannot guarantee the animals' safety in those areas without massive security investment.

A classic example is the DRC, which is the size of Western Europe with a population of fewer than 82 million. In stark contrast, crowded West Europe has 398 million people – almost five times as many as the DRC – so Africa's second largest country has vast swathes of wilderness. But to move endangered animals anywhere in the DRC's current turbulent political situation would be signing their death warrants. The last northern white rhino male died on 19 March 2018.

There are also huge tracts of land available in Angola, Zambia and Mozambique, but again, security will be a major issue. Mozambique has already lost its entire rhino population on two occasions. Can we risk that for a third time?

The answer obviously is no unless we have a highly trained militia guarding them around the clock. The costs will be enormous. A recurring nightmare for me is that conservation may one day culminate with the God-given right to see a rhino in the wild only being bestowed on the affluent few. Or no rhino left at all.

303

In South Africa, most of my waking hours are spent working with rhinos, whether in education or range expansion, which is one of my prime motives in the Loziba project.

But I also think constantly of the great apes, and my great friend John Kahekwa valiantly battling against overwhelming odds. Sadly, he lives in one of the most neglected and destitute countries in the world.

But there too, I have a plan. Granted, it's in embryo form, but it's still a plan.

Africa, I have always believed, will provide African solutions, and in this case the solution may well be through the embattled DRC's increasingly affluent neighbour, Rwanda.

Rwanda is turning out to be the most unlikely success story of the twenty-first century, emerging from the awful ashes of its horrific history like the proverbial flaming phoenix. It is a tiny country, but a giant in conservation and progress is happening at an inconceivable rate. I have met some of their wildlife officials and their cooperation and interest in eco-projects far, far outweighs anything else happening in Africa. Projects that take five years to complete in South Africa are springing up in Rwanda virtually overnight. There you can walk into the wildlife authority's offices, get permission and money to build a lodge within a day, and have it up and running within a few months. The state cooperation and desire to work up from ground zero is not found anywhere else on the continent. Perhaps even the world. As a result, investment is pouring in. Donors know they will get results.

Of particular interest to me is that at the moment they have an $18-million conservation fund obtained mostly from gorilla permits, with 10 per cent exclusively earmarked for community development in wildlife areas. This is tailormade for the grassroots work John is doing, and he is already considered a hero in Rwanda, despite coming from a different country. If a fraction of that community fund, say just 0.25 per cent, could spill over the border to the Pole Pole Foundation, it will be a massive boost. At the moment John doesn't even have a vehicle.

I know John, who is very proud of his Congolese roots, would rather have money coming direct to his country. But that seems unlikely at the moment.

Rwanda and the DRC are linked at the hip, separated by the Ruzizi River bridge, a mere stone's throw from John's home of Bakavu and Cyangugu on the Rwandan side. The path forward for Central Africa is through Rwanda, and I aim to make sure that John's magnificent community initiative will be on that thundering locomotive of progress.

So I am juggling several balls. Loziba is nearing fruition, but nothing happens fast in Africa. The Atherstone elephants are safe while we continue finding more homes.

But most inspiring to me is the runaway success of Rhino Art. It's astonishing to think that a fistful of crayons and scraps of A3 paper have changed the minds of hundreds of thousands of youngsters, from the source continent of Africa, to the market countries as far away as Vietnam. So far, we have reached nearly six hundred thousand children in thousands of schools. We are well on the way to my goal of a million.

How has this come about?

The answer may be found somewhere in the following account.

I ask you this . . . imagine you are in a ragged, dusty school, somewhere in the far outbacks of KwaZulu-Natal, where Richard Mabanga and I are holding a Rhino Art day. You are there, observing, watching, listening, thinking. You are breathing the ancient dust of Africa. You are part of the zeitgeist . . .

'*Haibo* – no!' more than five hundred children roar.

'It is true,' the man shouts back. He is dressed in a leopard-skin headband and *mbata* chest bib, animal-hide loincloth and *amashoba* white cow tails that bounce like frizzed ropes below his knees as he dances barefoot. One expects him to be waving an *assegai* or warrior's shield. Instead, he is holding a microphone that has been cranked up to full volume.

He points to me, standing next to him on a makeshift platform erected on the back of a truck. 'This man is actually a black man. My brother from across the Great Fish River. The land of the Xhosa.'

'*Haibo*,' laugh the children. The thought is too absurd to consider.

'Then why is his name *Nkunzi ayihlehli*, the bull who doesn't back down? Is that a white man's name?'

A hand goes up. Richard nods, waiting for the question.

'Then why does he look like a white man?'

'Ahh . . .' Richard Mabanga crouches, as if he is telling a story by a glowing fireside. 'You see, my brother, it goes like this. He was born black. But one day when he was a baby, it was so hot that his mother put him in a fridge while she was tilling the corn fields. Then she forgot about him. She forgot he was locked in a fridge. Can you believe that?'

'*Haibo!*' shout the children, shocked at such negligence.

'It is true. His mother only remembered she had a baby when it was time to feed him from her breast. But she couldn't find him. She looked everywhere before opening the fridge.

There he was. But he was so cold that the ice had turned him white. Today he is still white, even though he is no longer cold.'

The children hoot with delight.

Then I interrupt, speaking in Zulu. 'This man who tells this story,' I say pointing to Richard, 'well . . . his name is *Mahlembehlembe*. That's why he cannot stand still and talks all the time like a monkey chattering in a fruit tree. But what he says is true. I am his white brother.'

The children cheer. They cannot believe that this white man has a 'black' voice, speaking Zulu and even Seswati just as they do. They applaud as I switch to Xhosa, theatrically emphasising the tongue-clicks for which the language is famous.

Richard and I had arrived at the school an hour earlier. Even though we were expected, we respectfully paid our dues, asking the headmaster if we could address his pupils. We treat him as an *induna*, just as Ali Weakley taught me in my Xhosa classes when I was a boy, all those years ago. Next we are introduced to the teachers. We thank them for the work they do for our future generations.

Then we start setting up our mobile stage that we call the 'Rhino Rig', carried on a flatbed 1.5-ton pickup. But this is no bleak podium. It is strung with kite tails of colourful banners, bright flags, anti-poaching messages and an enormous ghetto-blaster sound system blaring *gumba-gumba* jive. The louder the better.

The kids watch from the classroom window. There is already a buzz of excitement. Something is in the air. They sense this is going to be a lot more fun than maths or science lessons. The festivity is palpable.

The bell rings and they burst out of the classrooms, crowding around the Rhino Rig, shouting, laughing and singing.

The show is on. We have forty-five minutes to enthral, enter-
tain and educate. No more. Anything longer may get boring.
Anything shorter may not punch the message through. The best
way to do it is through pure, unapologetic, over-the-top street
theatre. The actors are a black man dressed as a Zulu warrior
and a white man in Zulu migrant trousers sewn together with
more patches than Joseph's technicolour dream coat. They are
sending out a key message, subliminal or otherwise. It is a vital
message; an existential one. Yet it can be encapsulated in two
words: *Siyabathanda oBhejane* – we love rhino.

It is the message of our times.

Richard fuses entertainment and education so cunningly
that even I can barely tell the transition. The narrative is slipped
in effortlessly, without unnecessary verbal grease. He tells about
rhino horn being the same stuff as fingernails, yet men from
faraway places like Vietnam and China still think it can get
them a girlfriend. Can you believe that? No African would be
that stupid, he says. The children giggle hysterically at the
absurdity.

We then tell them how other people who also come from
faraway places pay a lot of money to see our animals. More
money than they can imagine.

'*Hau*,' they gasp. Paying money to see an animal?

'It is true,' says Richard. 'One animal can make the
community as much money in one year as a *spaza* shop. But
only if it is alive. Who pays to see a dead animal, rotten and
smelling with the flies buzzing around its body?'

We tell them about the game reserves, our God-given
privilege of having such beautiful animals roaming in our
country. We tell them that if we save the rhinos, instead of
killing them, there are going to be community projects in the

villages. Money from people coming to see rhinos will be put into buying more buses, more schools, more clinics. We tell them that God gave Africa the rhinos. Yet in almost all other countries, rhinos are no more. They have all been killed.

If we, the people of this land, let the rhinos be killed, we are not being true to the soul of Africa. It is our sacred covenant. We in South Africa are the last stand.

We tell them that the poachers will destroy all that. They will leave the people poor forever. Africa will be poorer as its spirit will be diminished if such an icon becomes nothing but a crayon picture like they have drawn for Rhino Art. If the rhino dies, a part of Africa's bones, the marrow of our land, will be lost forever.

As the show nears the end, Richard signals for silence. When the crowd is still, he roars into the microphone, 'What is the most dangerous animal on the planet?'

'*Bhubesi* – lion,' someone shouts.

Richard shakes his head.

'*Ndlovu* – elephant.'

Another shake of the head.

'*Ingwe* – leopard.'

Richard spreads his arms and shakes his head with a smile. Leopards seldom attack humans.

Then the kids rattle off a list: mambas, hippos, crocodiles . . . Richard keeps shaking his head.

Eventually someone understands.

'Humans.'

Richard nods and points to the kid who got the right answer. He or she gets a reward, a pen, a book or perhaps a soccer ball.

'No other animal kills as much as we do. It is humans who are killing the rhino. It is humans who must stop it.'

He then raises his voice.

'Do you know that you all have a rhino in your house?'

The children stare at him, uncomprehendingly.

'It is true. Every one of you have an *uBhejane* inside your home.'

'*Haibo*,' they shout back.

'Such an animal won't fit in my house,' yells one boy, and the children laugh.

Richard then pulls out a ten-rand note and waves it theatrically.

'When your mother sends you to buy bread at the *spaza* shop – what is the picture on your money? The ten rand she gives you?'

'Ahh,' the crowd sighs, '*uBhejane* – the black rhino.'

'Is that note of money in your home?' he shouts.

'*Yebo*,' the crowd responds.

'Remember that. Every time you buy bread, look at the rhino on your money.'

We pack up the mobile rig. The children are still dancing and singing. It is the Zulu way to express joy. We know we have reached them. We know they will remember our message. Whenever we return to a school, either to hand out prizes or to judge Rhino Art competitions, they rush to greet us shouting '*Siyabathanda oBhejane!*'

The meeting I have just described was a special one for me. I will never forget it. Not so much for the joy, vivacity and sheer electric energy, but what happened afterwards . . .

A boy comes up to me. Perhaps he is in his teens, but only just.

He is dressed in a dust-stained white shirt, grey shorts

hanging to his knees displaying scrawny legs with school regulation socks slipped down to his ankles. He shuffles nervously.

'*Baba*,' he says, 'I know where a bad man lives. A man who shoots rhinos.'

I look at him silently for a moment, then put my hand on his shoulder.

'Will you tell me where?'

'I cannot. My mother says I must not talk about it. He is a dangerous man. He has dangerous friends.'

I bend down to his level. 'I promise you no one will ever know what you say to me. It will be our secret. I will tell only one other man who will know what to do to stop this man. But no one will know it was you who first told me.'

I place my hand on my heart. 'I promise you that, *ndodana* – my son.'

The boy nods. I can see the uncertainty in his eyes. The fear radiates like a flashlight. At that instant, I understand the incalculable courage it has taken for him to come and speak to me. I am not ashamed to say my eyes are damp.

'He lives in the big house with the red wall in front.'

With that he was gone. I wanted to ask him how he knew the man was a poacher. I wanted to know what was the evidence. What more could he tell us.

But that would have been too much. The mere fact that he had come forward was an incredible breakthrough. Something Richard or I said during the stage show hit the target bullseye.

The spark of optimism that I long nurtured in my soul, often more in blind hope than belief, turned into a flame. Thanks to an *umfana*, a boy with the courage of a lion, I now knew the war could be won.

I passed the information on to a trusted friend in law enforcement, who then tipped off the police, saying it was confidential information so as to protect the schoolboy.

A week later, the man the youngster identified was stopped at a roadblock. The policeman, who had been briefed, was deliberately obtuse and the man responded angrily in kind. This was just the response the cop needed. The man was handcuffed, charged with obstruction and taken to his house with the red wall and the premises searched. An AK-47 was found hidden in a false floor, but no rhino horn. However, it was obvious the man was a poacher, and the fact that some of his 'friends' instantly fled the area was more than mere coincidence.

Having an assault rifle was enough to throw him in jail, although for how long you never knew with the local courts. Intelligence was able to match logistics to rhino crime scenes.

More crucial was the fact that a youngster watching one of our Rhino Art shows had delivered the goods. He was the first, but definitely not the last. We are now getting more and more people coming forward with information. Even teachers are whispering in our ears.

Most of what we get are 'cold' leads, where we hear of a group that has already poached in the area and left. Many are hesitant to provide names and addresses. But as any detective will tell, cold leads have a habit of suddenly becoming red-hot. The communities also know that anything given to us is in absolute secrecy.

Thanks to Rhino Art's vibrant mix of crude crayon drawings and street theatre, rural classrooms are now joining the front line. We never ask anyone to become informers, as that could be extremely dangerous. Instead, we hope the power of our message will persuade them to be eco-warriors.

We now know it is. That scrawny boy for me was the symbol of the grassroots fightback when we needed it most.

It has made my convoluted conservation journey from barefoot bumpkin to goat farmer, gorilla lover, elephant fixer and, above all rhino hugger – call it what you want – worthwhile.

I wouldn't have it any other way.

Glossary

Abakhwetha – Xhosa teenagers who have recently been through a circumcision ceremony

Amahashi – Zulu word for 'horses'

Amashoba – white cow tails worn on the lower leg in traditional Zulu warrior dress

Amasi – sour milk, part of the staple diet among rural Africans

Askari – young male elephant

Assegai – spear

Baba – Zulu word signifying respect for an older person, literally meaning 'father'

Bakkie – South African word for a pickup truck

Bhubesi – Zulu word for 'lion'

Biltong – South African jerky or dried meat, made from either beef or venison

Boerperd – a hardy South African horse, literally means 'farmer's horse'

Boet – Afrikaans word for 'brother'; also a generic term for 'friend'

BRREP – Black Rhino Range Expansion Project

Bwana – Swahili word for 'mister'

Chipembere – Shona word for 'rhinoceros'

CITES – Convention on International Trade in Endangered Species of Wild Fauna and Flora

DRC – Democratic Republic of the Congo

eGoli – Zulu word for Johannesburg

ERP – Elephants, Rhinos & People

Ezemvelo KZN Wildlife – the KwaZulu-Natal provincial wildlife authority

Gees – Afrikaans word for 'spirit'

Geza – Xhosa word for 'the naughty one'

Gumba-gumba – African jive music

Haibo – Zulu word for 'no'

Hau – Zulu expression of astonishment

ICCN – Congolese Institution for Nature Conservation

iMbongi – Zulu praise singer

Impi – Zulu warrior

Induna – Zulu and Xhosa word for 'headman' (plural: *izinduna/ iinduna*)

Ingwe – Zulu word for 'leopard'

Inkosi – Zulu and Xhosa word for 'chief' (plural: *amakhosi/ iinkosi*)

Isipho – Xhosa word for 'gift'

ITTO – International Tropical Timber Organization

Izintaba Zobombo – Lebombo Mountains, and the name of Kingsley Holgate's expedition along the range

Jambo – Swahili word for 'hello'

KBNP – Kahuzi Biega National Park

Kwedini – Xhosa word for 'boy'

Kwerekwere – Zulu slang for 'foreigner'

Linyeva likhishwa ngelinye – a warning to poachers by King Mswati of Swaziland: 'Put your hand in the fire and you will be burnt.'

Lobola – the price paid for a bride in Zulu and Xhosa cultures

Mahlembehlembe – nickname given to Richard Mabanga meaning 'all over the place'

Mbata – an animal-hide chest bib worn by Zulu warriors

Mblaselo – traditional Zulu trousers with colourful patches of cloth, beads and animal skins creating a uniquely African fashion style

Memsahib – Swahili word for 'madam'

Mfecane – Zulu word for 'the crushing', referring to the mass devastation caused by King Mzilikazi's foray into the South African interior

Moer – Afrikaans word for 'beat up' or 'wreck'

Muti – Zulu word for 'medicine'

Ndlovu – Zulu word for 'elephant'

Ndodana – Zulu word for 'son'

NGO – non-governmental organisation

Nkunzi ayihlehli – Zulu nickname for Grant Fowlds meaning 'the bull never backs down'

Nomabongo – Xhosa word for 'the proud lady'

Panga – African word for 'machete'

Pole pole – Swahili world for 'slowly-slowly' or 'gently does it'

Sangoma – traditional Zulu healer

SANParks – South African National Parks

Siyabathanda oBhejane – Zulu slogan meaning 'We love rhino'

Spaza – Zulu word for a general store stocking mainly groceries

Swaer – Afrikaans word for 'brother-in-law' but also generic for 'friend'

Thandi – Zulu/Xhosa word for 'love'

Themba – Zulu/Xhosa word for 'trust'

Thembi – Zulu/Xhosa word for 'hope'

uBhejane – Zulu word for 'black rhino'

Ulwaluko – Xhosa circumcision ritual symbolising the transition from youth to manhood

uMagundwane – Zulu word for 'rat'

Umfana – Zulu word for 'boy'

uMkhonto we Sizwe – the former armed wing of the African National Congress, literally 'Spear of the Nation'

Umphokoqo – dried, crumbly maize meal, staple diet for rural Xhosa

Veldschoen – Afrikaans word for 'bush shoes', usually made out of animal hide

Visa Volante – an entry visa needed for the Democratic Republic of the Congo

WJC – Wildlife Justice Commission based in The Hague

Yebo – Zulu word for 'yes'

Charities and NGOs

Project Rhino – www.projectrhinokzn.org
African Rhino Conservation Collaboration (ARCC) –
    www.arcc.org.za
Elephants, Rhinos & People (ERP) – www.erp.ngo
Chipembere Rhino Foundation – www.chipembere.org
Tusk – www.tusk.org
African Conservation Trust (ACT) – projectafrica.com
Freeland Foundation – www.freeland.org
Education for Nature – Vietnam (ENV) – envietnam.org
Wildlife Justice Commission (WJC) – wildlifejustice.org
12hours/Project Thorn – projectthorn.com
Kingsley Holgate Foundation and Rhino Art –
    www.kingsleyholgate.com
JuMu Rhino Fund – jumurhinofund.com
WildlifeDirect – wildlifedirect.org
Pole Pole Foundation – www.polepolefoundation.org
Jane Goodall Institute – www.janegoodall.org
One More Generation – onemoregeneration.org
WildAct Vietnam – www.wildact-vn.org
Traffic – www.traffic.org

BuyNoRhino – www.buynorhino.co.za

Wildlife Conservation Society (WCS) – www.wcs.org

African Parks – www.africanparks.org

Hoedspruit Endangered Species Centre (HESC) –
    hesc.co.za

Global Conservation Force – globalconservationforce.org

The Conservation Imperative – theconservationimperative.
    com

World Wildlife Fund (WWF) – www.worldwildlife.org

Black Rhino Range Expansion Project (BRREP) –
    http://www.wwf.org.za/our_work/initiatives/black_rhino_
    expansion.cfm

Wildlands Conservation Trust – wildtrust.co.za/wildlands

Index

3